Charles Robert Bree

A History of the Birds of Europe

Vol. 2

Charles Robert Bree

A History of the Birds of Europe
Vol. 2

ISBN/EAN: 9783337407643

Printed in Europe, USA, Canada, Australia, Japan

Cover: Foto ©berggeist007 / pixelio.de

More available books at **www.hansebooks.com**

A

HISTORY

OF THE

BIRDS OF EUROPE,

NOT OBSERVED IN THE BRITISH ISLES.

BY C. R. BREE, ESQ., M.D., F.L.S.,

Author of "Species Not Transmutable,"
Physician to the Essex and Colchester Hospital, &c., &c.

VOL. II.

———

LONDON:
GROOMBRIDGE AND SONS, PATERNOSTER ROW.
M DCCC LXIII.

CONTENTS

OF

THE SECOND VOLUME.

	PAGE
Guldenstadt's Redstart	1
Grey Redstart	6
Swedish Blue-throated Warbler	10
Ruby-throated Warbler	16
Thrush Nightingale	19
Barred Warbler	22
Ruppell's Warbler	26
Sub-alpine Warbler	29
Sardinian Warbler	33
Spectacled Warbler	38
Marmora's Warbler	42
Bonelli's Warbler	45
Olive Tree Warbler	49
Olivaceous Warbler	54
Vieillot's Willow Warbler	58
Black-throated Warbler	64
Marsh Warbler	70
Booted Reed Warbler	76
Aquatic Warbler	80
Moustached Warbler	85
Fantail Warbler	88
Cetti's Warbler	93
River Warbler	97

	PAGE
Pallas's Locustelle	101
Ruby-crowned Kinglet	108
Mountain Accentor	115
Black Wheatear	118
Russet Wheatear	123
Black-eared Wheatear	128
Pied Wheatear	133
Menetries' Wheatear	136
Yellow-headed Wagtail	138
Grey-headed Yellow Wagtail	143
Black-headed Yellow Wagtail	147
Sombre Wagtail	151
Red-throated Pipit	155
Water Pipit	164
Pennsylvanian Pipit	170
Tawny Pipit	175
Bifasciated Lark	179
Dupont's Lark	184
Desert Lark	188
Black Lark	192
Calandra Lark	195
Siberian Lark	200

BIRDS OF EUROPE,

NOT OBSERVED IN THE BRITISH ISLES.

INSECTIVORÆ.
Family SYLVIADÆ.
Genus SYLVIA. *(Latham.)*

Generic Characters.—Beak slender, compressed on the anterior half; the superior mandible sloped towards the point, the inferior mandible straight. Nostrils basal, lateral, ovoid, and half closed with a membrane. Tarsi longer than the middle toe; three toes in front and one behind, the external one united at its base to the middle; the claw of the hinder toe shorter than the toe, and curved. The first wing primary is very short, or absent, the second a little shorter than the third, or as long; the greater wing coverts much shorter than the primaries; tail extended, unequal, rounded, or square.

SECTION I.—RUTICILLÆ, (*Mühle.*) SYLVAINS, (*Temminck.*)

GULDENSTADT'S REDSTART.

Sylvia erythrogastra.

Sylvia erythrogastra,	Mühle; Monog. Europ. Sylviens.
Motacilla erythrogastra,	Guldenstadt; Nov. Comment., Petr. 19, 1775, p. 469.
" *aurorea var. ceraunia,*	Pallas; Zoog., 1, p. 478.
Lusciola erythrogastra,	Schlegel.
Rouge-queue de Güldenstädt,	Of the French.
Guldenstädt's Rothschwänzchen,	Of the Germans.

Specific Characters.—The tail unicolorous rust red, except the two middle feathers, which are brownish red; a white mirror from the third to the tenth primary. Length from the tip of the beak to the end of the tail seven inches. Length of wings four inches one line and a half, of tarsi one inch two lines, of middle toe eight lines, of hind toe five lines.

The group of birds of which the present is the first notice, is one of great interest to all who take the slightest pleasure in natural history. It contains all our summer Warblers—those harbingers of coming spring, which greet us with their merry or plaintive notes, or, as in the Nightingale, waken up the stillness of night with that full, rich, and beautiful song, which is unequalled by any music in the world. In all our summer walks or rides, however cultivated or barren the scenery, by wood or river, mountain or lake, we are sure either to see or hear some member of the family. We associate with our earliest days the croak of the Whitethroat, as flitting from branch to branch, or winding her way like a snake through the dense fence, she sought how often in vain to scare us from her nest. Who does not remember the "Renny Red-tail," and those old pollards in that quiet lane where the half-holidays of our youth were spent? or who will not always think with a corresponding touch of melancholy, of that low, soft, plaintive rebuke which the little green Willow Warbler poured into the ears of those who invaded its domed nest, so carefully hidden in the long grass; or the rich thrill of that saucy Blackcap, as he heralds the coming warmth of spring and summer by the song of triumph which announces his nuptial victory against all rivals? Then, again, there is the Grasshopper Warbler, with his invisible form and long sibilant note, and the Reed Wren, with his garrulous lecture, as he winds among the herbage by the river side, or the Sedge Warbler, as it sends forth in the still night its song of rivalry with the Nightingale. All these are salient beauties in that mental landscape which the naturalist often creates for himself, when the fortunes of life may have carried him among sterner and less poetical realities.

The *Sylviadæ* may be taken as typical of the *Insecticoræ*—their food being almost exclusively insects. But this is not quite true, for, notwithstanding the assertion of the late Mr. Yarrell to the contrary, the Willow Warbler will sometimes join in the more constant depredations of the Whitethroat.

Temminck divided the group into the *Riverains*, or those whose habits were aquatic; *Sylvains*, or those found more or less inhabiting woods; and *Muscivores*, or those which live principally upon flies, which they catch on leaves or on the wing. Count Von der Mühle has separated them into seven sections, which form, I think, a more natural division, and which I shall therefore adopt, giving at the same time that to which the bird is referred by Temminck, so as to keep up uniformity of arrangement. Count Mühle's sections are—

1. *Ruticillæ*, Rothlinge.—Redtails.

2. *Humicolæ*, Erdsänger.—Ground Warblers.
3. *Philacanthæ*, Grasmücken.—Hedge Warblers.
4. *Dumeticolæ*, Strauchsänger.—Bush Warblers.
5. *Phyllopseustæ*, Laubsänger.—Leaf Warblers.
6. *Polyglottæ*, Spötter.—Mockers.
7. *Calamodytæ*, Rohrsänger.—Reed Warblers.

The *Ruticillæ* or Red-tails, have long slender legs, tolerably long wings, and somewhat rounded tails, which are rusty or fox red, except the two middle ones. The males, and females, and young differ in plumage. Moult only once, and that during harvest.

Sylvia erythrogastra is an inhabitant of the Caucasus during the summer, migrating about the end of October to a warmer climate. It was first described by Güldenstädt in 1775, and was subsequently confounded with the *Motacilla aurorea* of Pallas, which is not, however, a European species, and is distinguished from it by the following characters, as given by Schlegel:—First, the two middle feathers in the tail of *aurorea* are black and not brown red. Second, the white mirror of the wings occupies the secondary quills to the eighteenth. Third, *aurorea* is altogether smaller, the total length being six inches; the wings three inches one line; tail two inches eight lines; tarsi eleven lines; middle toe six lines; and the hinder toe three lines.

According to Güldenstädt the manners and habits of this Warbler are similar to the rest of the family, approaching nearest to those of the Common Redstart. It lives on the shores of rivers, and is not a shy bird. It feeds on insects, and berries of the *Hippophæ rhamnoides*, in which also it builds its nest with grass. Nothing seems to be known about its eggs.

The figure which we have given of this bird is from Count Mühle's monograph, and is a copy of the specimen in the Leutchtenbergstein collection. The following is also Count Mühle's description:—"The old male in the spring has a very pretty and well-pencilled plumage. The whole surface of the head to the nape of the neck is of a brilliant satin white; the mantle, wings, flanks, cheeks, throat, and upper part of the chest, deep velvet black; the whole under part of the body, under wing coverts, rump, and tail, a lively fox red; on the deep black wings, and where the feathers are almost imperceptibly edged with a greyish tint, there is a great white mirror-like spot extending from the third to the tenth feathers. The secondaries have in the middle of the inner colours a white, though not an equally-conspicuous spot. The second primary is like the seventh, the third like the fifth. Beak black, with stiff black hairs round the angle of the mouth; throat yellow; iris brown; the tolerably long feet are black. After the autumn moult the plumage is much plainer.

The female has the colours of the under wing coverts and tail like the male, but paler; the whole body is brownish ash grey; on the under part of the body and about the sides lighter; under tail feathers rusty.

The unknown young birds we must suppose to be similar to other "Rothlings," clear spotted and streaked.

Figured by Heinrich Graf Von der Mühle, in his "Monog. der Europ. Sylvien."

INSECTIVORÆ.
Family *SYLVIADÆ.*
Genus Sylvia. *(Latham.)*

GREY REDSTART.

Sylvia cairii.

Ruticillæ carii,	Gerbe; Dict. Univ. D' hist. Nat. tom. xi. p. 259.
" *tithys,*	Brehm.
Erithacus carii,	Degland.

Specific Characters.—The primary and secondary quills fringed with ash-colour. General colour grey, with darker irregular markings. Dimensions of a male specimen sent to me by M. Ed. Verreaux, which is figured, length five inches and a half; from carpus to tip of wing three inches and three tenths; tarsi one inch; tail from vent two inches and a half.

This bird is exactly similar in plumage to the autumnal moult of the male *S. tithys,* the Black Redstart of the British lists. It affords a good example of what is termed a permanent variety or race; because it has been found by careful examination that it never assumes the dark breeding plumage of the latter bird. I have thought it right therefore to give a figure and notice of it in this work; more particularly as the Black

Redstart is a rare straggler in Britain, and not likely to be met with here in its moulting plumage.

Upon the question of *specific* difference, Degland observes: "M. Gerbe," who named the bird after the Abbé M. Caire, "writes me word that after much hesitation, and many objections made by his friend the Abbé, he had come to the conclusion that *R. carii* was only *R. tithys* in the autumn plumage." "But," continues M. Degland, "it is extraordinary that this hypothesis should be retained after the statement of M. Caire, corroborated by that of the Shepherds and Chasseurs, whom he consulted upon the subject, namely, that this bird breeds in its autumn plumage, which it never changes at any period of the year; that it only inhabits the highest regions of the Alps; that its song differs sensibly from that of *R. tithys;* and that it is only a bird of passage in the country inhabited by M. Caire, while the true *tithys* is sedentary there. Every research which has been made in the spring to find a bird in intermediate plumage has been in vain." And further M. Degland remarks, "This bird in its passage near Moustiers does not frequent the same places as *tithys;* it is only met in valleys, corn fields, meadows bordered with hedges, bushes, osiers, etc., into which it retreats at the slightest noise. On the contrary *tithys* always remains in rocky places, and upon old ruins. This difference in habit always prevented M. Caire, who gave me these details, from being mistaken, or from killing one instead of the other. His Redstart appears in the environs of Moustiers-Ste-Marie from the 5th. to the 15th. of April; that period past it may be looked for in vain, it is always very high up among the mountains. This year, M. Caire writes, they have been very abundant; he saw more than twenty during the 9th. and 10th. of April, and in the number he did not see a single *tithys*. The arrivals of *tithys*, which do not winter in the country, are always eight days later."

It is impossible, I think, to resist the force of these remarks. If not a species, *S. carii* is at least a permanent variety or race of a very interesting character, and quite as much entitled to specific distinction as the Shrikes described in a former number.

In estimating the value of colour as an element in determination of species, it cannot be too generally known that many of our most beautiful birds, such as the splendid Humming Birds, have in reality no colour at all, so far as this depends upon the deposit of pigment. The *effect* of colour is entirely produced on many feathers by delicate striæ, which decompose the light, the feather itself being colourless. This therefore is a *permanent* organic change in the structure of the bird, and not dependent upon those causes which influence the deposit of pigment. Colour then must be considered a valuable test of specific distinction when permanent. In the present instance, that of *S. carii*, the peculiarity appears to be in the incapability of producing breeding plumage feathers. This is propagated from one to another, like many other abnormal peculiarities in all classes of animals, and the result is a permanent variety or race, which the sagacity of Temminck and others led them to indicate as distinct from *specific* difference.

The *Sylvia carii*, or, as I have taken the liberty of calling it, the Grey Redstart, inhabits the top of the Basses-Alpes, where it breeds in old isolated chalets or huts. "It lays two sets of eggs, one at the end of April in more temperate regions, when the high mountains are covered with snow; the other at the summit of these mountains near the eternal snows, where the Alpine Accentor and Snowfinch breed, and where one never sees by any chance a specimen of *tithys*."

Its nest is composed of dry blades of grass, and is lined with a quantity of feathers, in which is laid four or five white eggs, but of a paler hue than those of *tithys*, and slightly shaded with blue.

The description given by M. Degland, who appears to have paid much attention to this interesting race, is exactly like the skins sent me by M. Verreaux.

The male has the entire plumage of an ashy brown or grey; the under parts slightly clearer, with a russet shade on the top and front of the head; the space between the beak, the eyes, and parotid region brown; free edge of eyelids grey; the fringe of the secondary quills of the wings much less than that of *tithys*, and grey; all the primaries bordered with clear ash-colour; the upper tail coverts of a lively russet; the under tail coverts of a reddish white; all the feathers of the tail red except the two middle ones, which retain the general colour of the plumage; beak and tarsi black.

The female has the plumage slightly lighter than the male, and shaded with russet on the chest; no brown between the beak and eyes; throat reddish; the rest as in the male.

S. carii differs from *tithys*, according to Degland, by; first, absence of black from plumage; second, by the edges of secondaries which, instead of being white, and so large as to form a sort of mirror on the folded wing, are grey, and scarcely visible.

INSECTIVORÆ.

Family *SYLVIADÆ.*

Genus SYLVIA. *(Latham.)*

SECTION II.—HUMICOLÆ, (*Mühle.*) Ground Warblers.
SYLVAINS, (*Temminck.*)

SWEDISH BLUE-THROATED WARBLER.

Sylvia cærulecula.

Motacilla suecica,	LINNÆUS S. N. 12th. Edit. 1766, tom. i., p. 336.
" "	GMELIN; Syst. 1788, tom. i., p. 989.
" *cærulecula,*	PALLAS; Zoog. 1811, p. 480.
Sylvia suecica,	LATHAM; Ind. 1790, tom. ii., p. 521.
Lusciola suecica,	KEYSERLING AND BLASIUS; Die Wirbelt, 1840.
" *cyanecula orientalis,*	SCHLEGEL; Revue, 1844.
Ostlicher Blaukehlchen,	OF THE GERMANS.
Gorge-bleu orientale,	OF THE FRENCH.

Specific Characters.—Throat and neck blue, with a rusty red spot in the centre; tail feathers dark brown; the basal half bright red, except the two middle ones. Length of adult male sent me

by M. Verreaux, five inches and a half; from carpus to tip three inches; tarsus thirteen lines; tail two inches; beak, the lower mandible six tenths, and the upper eleven twentieths of an inch.

The bird figured by Mr. Yarrell as *Phænicura suecica* is the *S. cyanecula* of Brisson, Meyer, and Schinz, and the variety *B. suecica* of Gmelin and Latham. The real *Motacilla suecica* of Linnæus is, according to Degland, the bird which I have figured, and which in a dozen skins sent me by M. J. Verreaux, has the red mark in the centre of the blue more or less developed. It is most probably one of two varieties, and it is very difficult to decide to which of the two the typical form can be referred.

Yarrell says that the white spot of *Phænicura cyanecula*, the bird which he has described and figured as the original *Motacilla suecica* of Linnæus, is sometimes seen in very old males red. M. Temminck also describes the red-spotted variety as a permanent race, only occurring in Russia and Siberia. Degland, on the contrary, says that the Russian specimens have the white spot as well as the red one, and some have the spot partly red and partly white, as though the two races had crossed; and the same has been remarked by Mr. Hardy of specimens sent to him from the Nile.

Professor Allman has received during the present year a series of *S. cærulecula*, sent to him for the Museum of Edinburgh, from Heligoland, by Mr. H. Gätke. They are adult male and female, and three young ones. In one of the latter there is a clear white spot, with a tinge of red appearing round it. The occurrence of this variety in Heligoland, Norway, and Sweden generally, strengthens Degland's opinion that it is the genuine *Motacilla suecica* of Linnæus. It breeds regularly in the island, and the white-spotted variety is not known there.

M. J. Verreaux says that the red-spotted variety is constant, and the Indian skins which I have examined have this mark in all the specimens. Schlegel has called the variety *Lusciola cyanecula orientalis;* but Mr. Blyth referred the bird he has described in India to the white-spotted variety, and considers it, like most other ornithologists, as the *Motacilla suecica* of Linnæus. All this confusion might have been avoided had naturalists merely described the two birds as varieties of each other, which it appears to me they are undoubtedly, although the constant character of the white or red spots evidently entitles them to be considered as separate *races* of the same original stock. As only one bird has been figured and described as occasionally found in England, I take the opportunity of introducing the other variety in this work.

We have in this bird one of the many illustrations which the modern system of nomenclature affords us of the impropriety of tampering with generic names. Linnæus placed it in the genus *Motacilla*, to which it is in its habits closely allied. Latham subsequently removed it into his *Sylvia*, where it ought to have finally rested. In accordance however with what is termed the "system of nomenclature adapted to the progress of Science," we find it changed by one writer to *Lusciola*, by another to *Cyanecula*. Selby placed it with the Redstarts in the genus *Phœnicura*, Eyton in that of *Ficedula*, Degland in that of *Erithacus*, and Hodgson in that of *Calliope*. I trust I need only mention these facts to condemn a system false in principle and injurious to the progress of Natural Science; for we must always bear in mind that the general reader and

WEDGE BLUE-THROATED WARBLER.
RUBY-THROATED WARBLER.

student, for whom all systems are formed, is apt to draw conclusions upon the subject by no means flattering to the judgment or acuteness of the writer.

The *Sylvia cærulecula* is found inhabiting Russia and Siberia, Lapland, Norway, and Sweden. It occurs in the East Indies, and the species is identical with that of Europe. Degland notices a male having been killed the end of April, 1836, near Douai, and others he says have been captured at Burgoyne and in Picardy. M. Malherbe possesses two specimens, which were killed near Metz, and its plentiful occurrence in Sweden and Norway has been recorded by M. M. De Lamotte and De Cosette, in 1829. Count Mühle notices the appearance of isolated individuals in Greece. They are found in the autumn, and appear to be migrating from the north-east to the south-west. Only the Russian, or more properly the Norwegian variety *(cærulecula)* occurs.

The following extracts from the Journal of the Asiatic Society and the Madras Journal, give an interesting account of the habits of the "Blue-breast," by Mr. Blyth and Mr. Jerdon. From what we have said it clearly does not signify much which of the two races it refers to, as their habits are most probably precisely the same.

Mr. Blyth says, "The Blue-breast affects the open country where there are no trees, and especially reedy places or plantations of sugar-cane, or growing corn or high grass, or ground covered with the broad leaves of cucurbitaceous plants; and there they are generally seen on the ground, running with alternate steps, like a Pipit or Wagtail, and occasionally spreading wide the tail, displaying its rufous base to advantage; seldom perching, but flitting before you as you advance, and disappearing among the low cover, but soon coming forth when all is still, yet without absolutely quitting the shelter of the herbage by going more than a few paces from it. In Lower Bengal these birds are extremely common in suitable situations."

Mr. Jerdon remarks, "In the Dukhun this is far from being common, and is only found during the cold season—from October to March. It frequents thick hedges, gardens, sugar-cane fields, and long grass or weeds in beds of tanks, etc., occasionally coming close to houses. It feeds on the ground, on which it runs along, picking up various insects, and does not return so quickly to its perch, neither has it that peculiar quivering of the tail, as the Indian Redstart, though while feeding on the ground it occasionally jerks it up. It generally tries to conceal itself among the bushes it frequents when observed."

The male in breeding plumage has the upper parts and wings greyish brown; rump reddish; throat, from base of lower mandible to crop, bright blue, with a rusty red mark, or, as Temminck calls it, "mirror" in the centre; the lowest blue feathers are edged with grey, and immediately below there is a broad black band, succeeded by one of a bright russet; abdomen and flanks are of a dirty white, with dusky and irregular reddish markings; under tail coverts yellowish red; tail dark brown, the basal half of each feather, except the two middle ones, bright russet red; beak black; iris and legs brown.

The female differs from the male in having the throat whitish, margined only with blue; the colours across the chest quite distinct; the abdomen is more uniformly yellowish white.

In the young, before the first moult, the plumage is

slightly browner above than that of the adult, with the throat whitish, mottled with black spots here and there. In older birds the black spots form a ring more or less broad, with a white space or mirror in the centre.

INSECTIVORÆ.
Family *SYLVIADÆ.*
Genus SYLVIA. *(Latham.)*

RUBY-THROATED WARBLER.

Sylvia calliope.

Motacilla calliope,	PALLAS; Voy., 1776.
" "	GMELIN; Syst. N., 1788.
Turdus camtschatkensis,	GMELIN; S. N. L.
" *calliope,*	LATHAM; 1790.
Accentor calliope,	TEMMINCK; 1835.
Calliope lathami,	GOULD.
" *camtschatkensis,*	BONAPARTE; 1838. BLYTH.
" "	HODGSON.
Lusciola (melodes) calliope,	KEYSERLING AND BLASIUS; 1840.
" " "	SCHLEGEL; 1844.
Cyanecula calliope,	GRAY; Gen. Birds.
Sylvia calliope,	MÜHLE; 1856.
Gorge en feu calliope,	OF THE FRENCH.
Feuerkehlchen Sänger,	OF THE GERMANS.
Gunpigera,	BENGAL.
Gangular,	NEPAL.
Nogotto,	JAPAN.

Specific Characters.—Upper parts olive green; primaries and tail hair brown. Under parts whitish, mottled with olive; throat and neck of a clear brilliant vermilion red, lighter in the female. Length of male six inches and a half; carpus to tip three inches; tarsus one inch and three fifths; middle toe one inch and one fifth. Female rather less.

This beautiful Warbler is an inhabitant of India, the Philippine Islands, Japan, and other parts of the eastern world. It is also found in Siberia, Kamtschatka, and occasionally in Russia and the Crimea, in consequence of which it has been introduced into the European fauna. According to Blyth, it is common in Lower Bengal during the cold season. In Eastern Siberia it is found plentifully at Jenisei, Selenga, and Angara, where it arrives in May and disappears in September.

It was placed by Temminck among the Accentors, but more naturally, I think, by Mühle among the *Sylvia*, and near the Blue-throated Warbler. Its long synonymy will shew the contest which has arisen in men's minds as to the right designation of this interesting songster. But if a bird which in structure and many parts of its plumage closely resembles the Nightingale, which, as we are told by Pallas, frequents willow bushes, from the top of which, emulating the queen of song, it sends forth its glorious notes at sunrise, noon, or midnight alike; if, I say, such a bird does not deserve to be ranked among the Warblers, and to stand high in the well-marked family of *Sylvia*, I cannot understand how we are to make natural affinities the bonds or links of classification.

Of the nidification of *S. calliope* we know very little. Pallas tells us that it builds a careless nest, and that its eggs are greenish in colour, and that its call-note may be heard as it flies. It is entirely insectivorous.

Male.—The whole upper part of the body is olive green; top of the head, primaries, and tail, umber brown tinged with olive; a clear white streak over the eyes, and another larger one extending from the base of the lower mandible to the middle of the side of

the neck. The lore and base of the inferior mandible
deep black; the throat and upper part of the neck
brilliant vermilion red, bordered with dark grey; middle
of abdomen and under tail coverts bluish white;
crop and flanks greenish grey or olive brown; middle
tail feathers the longest, and rounded; the side feathers,
particularly the outermost ones, pointed. Beak and iris
brown; feet brown. In autumn the splendid red
throat is covered with white feathers.

The female has the upper parts like the male, but
the red on the throat is much lighter, and in some
specimens it is described as being more of a rosy
tint. In the East Indian specimen before me, sent
by M. Verreaux, the vermilion red is still retained,
though lighter, and with a whitish spot in the centre.
In my specimen the lore is black, and the vermilion
red is surrounded by a bright grey border, becoming
black as it comes in contact with the white line which
extends from the base of the mandibles.

The young male is described by Degland as similar
to the adult, with the throat and upper part of neck
of a clear rose or yellowish red.

My figure of this bird is a male in breeding plumage,
from an Indian specimen sent to me by M. Verreaux.
It has also been figured by Gould, in "B. of E."

INSECTIVORÆ.

Family SYLVIADÆ.

Genus SYLVIA. *(Latham.)*

THRUSH NIGHTINGALE.

Sylvia philomela.

Luscinia major,	BRISSON; Ornith., 1760.
" "	NAUMANN. GOULD.
Motacilla luscinia major, var.,	GMELIN; Syst., 1788.
Sylvia philomela,	BECHSTEIN; 1810.
" "	TEMMINCK; Manual, 1820.
" "	SCHINZ; Europ. Fauna, 1840.
Lusciola philomela,	KEYSERLING AND BLASIUS.
" "	SCHLEGEL.
Bec-fin or Rossignol philomele, or Grand Rossignol,	OF THE FRENCH.
Sprosser Nachtigall,	OF THE GERMANS.
Rossignolo Forestiero,	OF SAVI.

Specific Characters.—Plumage above of a sombre brown. The first primary nearly obsolete, the second nearly as long as the third, and longer than the fourth. Length of male and female seven inches; from carpus to tip three inches and a half; tail from vent three inches; beak from gape seven tenths of an inch.

THE Thrush Nightingale is an inhabitant of the eastern parts of Europe. It is found in the south of Sweden—in Pomerania and Finland, in the south of Germany and Switzerland, Poland, Hungary, and

Dalmatia. According to Temminck it is also found in Spain, but this is doubted by Count Mühle. It ranges south as far as the Volga, the Caucasus, Egypt, and Persia. It is rare in France, though M. Gerbe records the appearance of two specimens in the neighbourhood of Paris, in September, 1847. It does not occur in Holland.

Count Mühle informs us that it appears later than the Nightingale, but that it chooses the same localities, preferring, however, the neighbourhood of water and marshes. It likes to select its dwelling in the deep-lying thickets among the cultivated islands on the large rivers. In Germany it especially frequents the shores of the Don, Oder, Elbe, and their tributaries, but is rarely found on the Rhine.

In its habits the Thrush Nightingale appears more impetuous and not so graceful as its congener; its song is deeper and louder, and by some not thought so pleasing. It builds generally on stumps of trees. It lays five or six eggs, which, as will be seen by the figure of a specimen sent me by M. Verreaux, are very similar to those of our well-known species. Count Mühle says the egg is generally darkly spotted, which is not however mentioned by Temminck, and denied by Degland. All the specimens sent me by M. Verreaux are deep olive, like that figured; one lighter in colour, but none of them with any spots.

The plumage is so like that of the Common Nightingale, that it is not necessary to give any lengthened detail. The male and female have the upper parts of a dull grey brown; clear grey, tinted with darker on the chest; tail less brightly marked with russet than in the Common Nightingale; throat white, surrounded by dark grey; feet brown.

In the young before the first moult the upper parts are bright clear russet brown; the head, scapularies, and wing coverts, thickly mottled with light chesnut. All the under parts mottled with dark grey and dirty white, with a shade of yellow; tail rich chesnut brown; feet and legs light brown; beak rather darker.

The figure is from an adult male sent me by M. Verreaux. It is also figured by Gould, "B. of E."

INSECTIVORÆ.
Family *SYLVIADÆ*.
Genus SYLVIA. *(Latham.)*

SECTION III.—PHILACANTHÆ. GRASMÜCKEN.
Hedge Warblers, (*Mühle.*) SYLVAINS, (*Temminck.*)

BARRED WARBLER.

Sylvia nisoria.

Sylvia nisoria,	BECHSTEIN. MEYER AND WOLFF.
" "	TEMMINCK. VIEILLOT. CUVIER.
" "	KEYSERLING AND BLASIUS.
" "	SCHINZ. SCHLEGEL. DEGLAND.
Nisoria undata,	BONAPARTE. NAUMANN.
Curruca undata,	GERBE; Dict., 1848.
Bec-fin rayé, or	
Fauvette epervière,	OF THE FRENCH.
Sperber Grasmücke,	GERMAN.
Celega padovana,	SAVI.

Specific Characters.—Secondaries fringed with light grey; the middle tail quills and under coverts broadly bordered with white; the lateral tail quills with a white spot at the extremities and inner borders. Length of adult female from M. Verreaux, which is figured, seven inches; carpus to tip three inches and a half; tail two inches and a half; tarsus one inch.

THIS is one of the largest species of European *Sylviadæ*. It inhabits particularly the north and

eastern parts of Europe. It is found on the shores of the Mediterranean, and thence to Sweden and Norway, the north of Germany and some parts of Russia, and in Hungary. It is more rare in Austria, but is found in Lombardy, Piedmont, Central Italy, and the coast of Barbary. Count Mühle doubts if it occurs in the Pyrenees. According to Temminck, it is found accidentally in Provence, and during its passage in Tuscany; less rare in the Levant, and common in the neighbourhood of Vienna.

The Barred Warbler belongs to the same division as the Whitethroats, and, like them, though of considerable size and somewhat clumsy appearance, it is swift and active. It lives generally very much concealed, and is not, therefore, so often observed as the other members of the group. It has a particular predilection for thorny bushes. Avoiding mountainous districts, it is found in field hedges and young thickets, particularly where blackthorn and whitethorn abound. In spring it is observed in woods on high trees about the period of migration, getting again into the thickets in the autumn. It is a restless bird, never known to sit still, hopping about from branch to branch, and gliding along the fence with considerable rapidity. When met with suddenly, it raises up the feathers on the top of its head, like the Common Whitethroat, jerks up its tail, and utters a harsh cry.

Count Mühle does not give our bird a very high character. He says it is not only a very restless, but also a very quarrelsome and jealous fellow, driving away all other birds out of its hunting district; and while the lady bird is performing the duties of incubation, her lord is assiduously engaged in driving off all disturbers of the peace.

The note of *S. nisoria* is strong and melodious, and it sings from early morning till late in the evening a song not inferior to that of the Garden Warbler, which it somewhat resembles. It sings frequently while flying, and may often be seen rising up several yards into the air, and then falling down like a shot upon another tree or bush, alternately flying and fluttering.

In autumn the young males may be heard like the other Hedge Warblers, snapping and croaking to all comers.

The Barred Warbler, according to Mühle, builds in the beginning of May in thick thorn bushes, at a height of two to four feet, a slight half-globular-shaped nest. It is made very loose, with dry stalks of plants, small straws interwoven with spiders' and caterpillars' webs, and lined inside with horse-hair. It lays four to six eggs, grey greenish or yellowish grey, with bright ash grey or pale brown spots. They only breed once a year, and the male and female sit on the eggs alternately.

The male has the whole upper part of the body clear dark grey, usually with a rusty yellowish tint. The greater and lesser wing coverts and upper tail coverts edged with white, more feeble and contracted from the third to the fifth primary; the third primary longest, the second almost as long; tail dark ash grey, first feathers with outer border whitish; the middle quill and second quill have at the end a broad wedge-shaped white spot; the third and fourth an oblique deep edge of white. Under parts of body greyish white, and flanks darker, with dark grey wavy lines, especially well marked on the under tail coverts, taking there a lanceolate form. In the young birds these wavy lines are indistinct, but as they increase in age, especially the males, they become more numerous and darker. Beak slightly hooked at the tip; iris deep yellow; feet yellowish grey.

In the female the colour is duller above, and the chest and flanks tinted with russet; the white spots at the end of the tail are smaller and not so distinct.

The young before the first moult, according to Temminck, are of a uniform grey. Vieillot and Mühle say they are covered with crescentic spots, which are greyish brown on the neck and throat, chest and flanks.

After the first moult they have the upper parts grey, with indistinct bands of a russet white; under parts white, except the flanks, which are very slightly marked with grey.

The bird figured is an adult female sent to me by M. Verreaux. The egg also came from the same gentleman.

This bird has also been figured by Roux, Ornith. Prov., pl. 222, (male;) Gould, B. of E.; and Naumann, Taf., 76.

As Gould and Roux have figured males I have thought it best to give a drawing of the female, though males are figured in this work as a general rule.

INSECTIVORÆ.

Family *SYLVIADÆ.*

Genus SYLVIA. *(Latham.)*

SECTION IV.—DUMETICOLÆ. STRAUCHSANGER.
Bush Warblers, *(Mühle.)* SYLVAINS, *(Temminck.)*

RUPPELL'S WARBLER.

Sylvia ruppellii.

Sylvia rüppellii,	TEMMINCK; Manual, 1835.
" "	KEYSERLING AND BLASIUS, 1840.
" "	SCHINZ; Europ. Fauna, 1840.
" "	SCHLEGEL; Revue, 1844.
Curruca rüppellii,	BONAPARTE, 1838.
" "	GERBE; Dict., 1848.
Bec-fin Rüppell,	OF THE FRENCH.
Rüppell's Sänger or Grasmücke,	OF THE GERMANS.

Specific Characters.—Above ash grey; tail black, the outermost quill feather clear white; on the two following, the tip and a wedge-shaped spot on the inner web, white; on the fourth and fifth also a small white spot on the tip. Length about five inches and a half.—MÜHLE.

THIS Warbler is an Asiatic species, inhabiting especially the borders of the Red Sea and the Nile. It was introduced into the European fauna by Temminck in

the last edition of his " Manual" in 1835. Its European locality has hitherto been confined to Greece, where it was observed, though rarely, by Count Mühle. According to Lindermayer it occurs in the bushy ravines of the Attic Mountain range, but Count Mühle found it only in the Morea. The single specimen he captured, he informs us, was "sitting on the outstretching branch of a bush in the hollow of a rocky ravine."

It appears in Greece in May, and leaves in August. It does not appear to be so sprightly or quick in its movements as its congener, the Dartford Warbler. It will sit on the end of a branch with "hanging tail" while guns are fired in the neighbourhood, without being alarmed. Count Mühle adds nothing about its song, and says that its nidification and propagation is one of the points in its natural history still to be elucidated. Thienemann says the nest is cup-shaped, somewhat scantily and loosely built of dry stems of plants, dry leaves, strips of bark and vines loosely lined inside with softer materials. The ground-colour of the eggs is milk or yellowish white, with delicate pale green and grey green spots, which form a narrow ring near the base.

I take the description from Count Mühle:—The whole upper part of the body is ash blue grey; the wings are brownish black; the greater wing coverts, as well as the hind feathers of the wing are bordered with a whitish circle; the primaries are marked externally with a whitish border; first very short, second and fifth of same length, the third the longest, very nearly the same length as the fourth. The slender black tail is rounded; the outside feathers entirely white, except at the root, which is blackish; the shafts white. Under parts white, going off into ash grey in the flanks. Bill small, and contracted at the sides, curved from the middle, and hollowed out at the tip—in colour, horn black, the lower mandible yellow; iris nut brown; superciliary feathers white; the naked eyelids deep cinnamon red; the feet strong and horn yellow.

The male has the crown of the head velvet black, which extends to the lore and under the eye; auricular orifice grey; cheeks dark ash, with a white band or moustache, which, from the angle of the mouth, extends along the sides of the neck, and encloses the black of the throat; the white of the under parts is delicately tinged with rose-colour.

The female has the top of the head, throat, and breast, dark ash grey, and the white parts are not tinged with rose.

In the young the grey of the upper part of the body is without spot, and duller; the throat is whitish.

According to Temminck the black feathers of the head and throat after moulting appear as white plumage, which is by degrees rubbed off.

The figures of this bird are by permission taken from Mr. Gould's B. of E. The egg is from Thienemann.

Figured also by Temminck and Laug, pl. color, 245, f. 1, (male.)

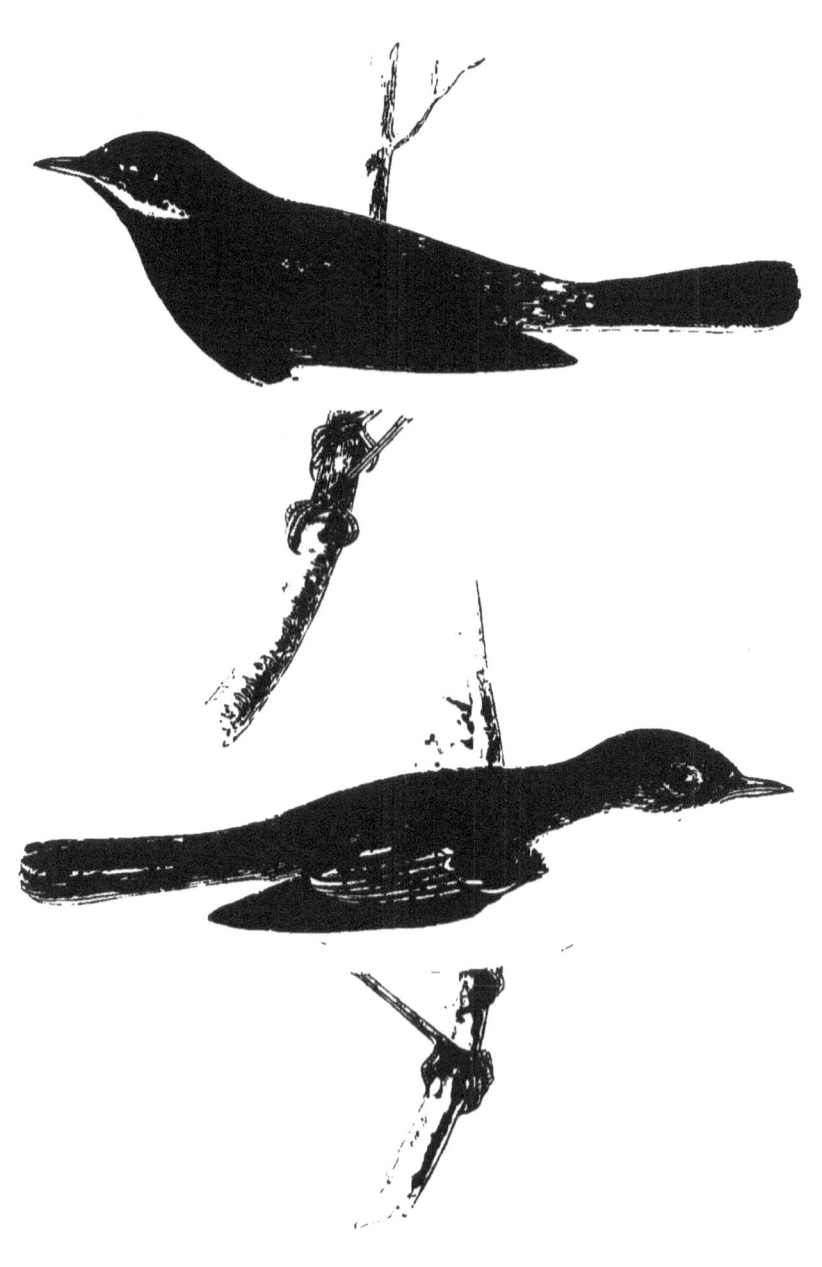

INSECTIVORÆ.
Family SYLVIADÆ.
Genus SYLVIA. *(Latham.)*

SUB-ALPINE WARBLER.

Sylvia sub-alpina.

Sylvia sub-alpina,		BONELLI.
"	"	TEMMINCK; Man., 1, p. 214.
"	"	BONAPARTE; Consp. Gen. Av.
"	"	SCHLEGEL, 1844. MÜHLE.
"	*passerina,*	GMELIN; Syst., 1, p. 954.
"	"	LATHAM; Ind., 2, p. 508.
"	"	TEMMINCK; Man., 1835, p. 131.
"	"	DEGLAND, 1844.
"	*leucopogon,*	MEYER AND WOLFF; Orn. Tasch., 3, p. 91.
"	"	SAVI; Orn. Tosc., vol. 1, p. 257.
"	"	SCHINZ; Fauna Europ.
"	*bonelli,*	KEYSERLING AND BLASIUS.
"	*mystacea,*	MENETRIES; Cat., p. 34.
Curruca passerina,		Z. GERBE; Dict. un d' Hist. Nat., 1848.
Fauvette sub-alpine, or *Bec-fin Passerinette,*		OF THE FRENCH.
Weissbärtiger Strauchsänger,		OF THE GERMANS.
Sterpazzolina,		OF THE ITALIANS.
Passerine Warbler,		LATHAM.

Specific Characters.—The inner barb of the primaries and secondaries edged with white in the male, unicolorous olive brown in the female. Edge of carpal joint in both sexes white. Length of an adult male from M. E. Verreaux, four inches and a half; from carpus to tip two inches and two tenths; tail two inches and three tenths; tarsus four fifths of an inch. Female about same size.

This bird was described by Temminck in the first edition of the "Manual" as a new species. Savi and Roux, however, clearly demonstrated that it was not specifically distinct from *Sylvia passerina*, described in the third volume (edition 1835) of the "Manual," but only that bird in the breeding plumage. Schlegel and Mühle have consequently sunk the name of *passerina*, and adopted that of *sub-alpina*, given to it by Bonelli. To add to the confusion Meyer and Wolff have also described one of the plumages of this bird as a distinct species, *Sylvia leucopogon*. It will be necessary, therefore, to bear in mind that *S. passerina* of Temminck, and *S. leucopogon* of Meyer and Wolff, are both comprised in the *Sylvia sub-alpina* of this notice.

The Sub-alpine Warbler has a wide range both in Europe and Africa. It lives along the whole coast of the Mediterranean, is abundant in Algeria and Egypt, and is found in Sardinia, Italy, Dalmatia, Silesia, and in the Steppes of New Russia and the Ghouriel. It also occurs plentifully on the borders of the Caspian Sea. In the South of Europe it generally appears with the other migratory Warblers in the beginning of April. In Greece Count Mühle informs us it is seen about the middle of March, in the low bushes and shrubs along the dried-up river-beds in the mountainous part of the country. In Italy it occupies similar localities, in company with the Common Whitethroat, and leaves in September or beginning of October.

According to Savi it is not often seen consorting with Blackcaps or Garden Warblers, although its song is somewhat of the same character. When the male wishes to sing he glides out of the bushes, and, perching on a neighbouring branch, sends forth his clear agreeable melody. When he has finished, or is disturbed, he glides again into the fence, and only makes his whereabouts cognizable by his frequent call-note, similar to that of the Common Wren. These habits fortunately render it a very difficult bird to capture.

It builds twice a year; its nest is globe-shaped, placed in thick bushes from three to five feet from the ground, formed outwardly of straw and withered grass, and within of delicate small roots, rarely lined with wool, and interwoven with the webs of spiders and caterpillars. It lays four or five eggs, roundish, greenish white, sprinkled with small brown spots, which are most numerous and large at the broadest end. The male relieves the female in incubation during the middle of the day.—Mühle.

According to M. Moquin-Tandon, as quoted by Degland, M. M. Webb and Berthelot brought from the Canary Islands eggs of this species, which were exactly similar to those which he took in the neighbourhood of Montpellier.

The male in breeding plumage has the head, nape, and scapularies of a bluish lead-colour; upper parts of the wing and tail olive brown. The throat, crop, and flanks russet red, more or less strongly marked; middle of the belly whitish, mottled with bluish spots; a white line or moustache from the gape separates the grey of the nape from the red of the throat and crop. Under tail coverts white, shaded with russet; two outer feathers of the tail white on each side above and inside for three parts of their length, the two following only tipped with white; beak brown, reddish at the base below; iris yellow; legs and feet flesh-coloured. In autumn the upper parts are grey, more or less tinted with olive or russet; inferior parts of a less bright red, clearer on the flanks, and the abdomen whiter.

The female has the upper parts uniformly olive brown, with a tinge of bluish grey about the head and nape; the under parts much less red than in the male, but in my specimen the colour which is a faintly reddish white, is more uniformly dispersed.

The young before the first moult have the superior parts, according to Degland, (whose descriptions are in general most accurate,) of a reddish ash; the inferior parts reddish, or a clear brown, with the middle of the abdomen white. Wings brown, all the coverts being broadly bordered with reddish; tail feathers brown, fringed with reddish ash, the external feather of each side bordered and terminated with whitish ash. In a young male bird sent me by M. Verreaux, the colours are very similar to, but fainter than those of the adult male; the abdomen is more mottled. In none of the specimens is the abdomen of a pure white.

Figured by Temminck and Laugier, pl. col. 251, f. 2 and 3, (male and female;) Roux, Ornith. Prov., pl. 218, f. 1 and 2, (adults;) Gould, B. of E.

My figures of this bird are from specimens sent me by M. Verreaux.

INSECTIVORÆ.
Family *SYLVIADÆ*.
Genus SYLVIA. *(Latham.)*

SARDINIAN WARBLER.

Sylvia melanocephala.

Sylvia melanocephala,	LATHAM; 1790.
" "	TEMMINCK; Man., 1820.
" "	BONAPARTE; (B. of E.,) 1838.
" "	KEYSERLING AND BLASIUS; 1840.
" "	SCHINZ; Europ. Faun., 1840.
" "	SCHLEGEL; Revue, 1844.
" "	DEGLAND; 1844.
" "	MÜHLE; 1856.
Motacilla melanocephala,	GMELIN; Syst., 1788.
Sylvia ruscicola,	VIEILLOT; Dict., 1817.
Curruca melanocephala,	LESSON; Ornith., 1831.
" "	Z. GERBE; 1848.
Pyrophthalma melanocephala,	BONAPARTE; Consp. Gen. Av.
Bec fin mélanocéphale, or Fauvette des fragons,	OF THE FRENCH.
Schwarzköpfiger Strauch-sanger,	OF THE GERMANS.
Occhiocotto,	OF THE ITALIANS.

Specific Characters.—Throat white; head black in the male, greyish black in the female; the secondary quills fringed with russet grey; the two external tail quills white on the outer side, with a large spot of the same colour at the extremity of the internal barb; the other tail feathers white only at the tip.

Length of an adult male sent me by M. E. Verreaux, five inches and a half; from carpus to tip two inches and three tenths; tail two inches and a half; tarsus seven tenths of an inch.

The Sardinian Warbler inhabits the south of Europe, the Canary Islands, Egypt, and, according to Degland, Asia Minor. It is also included in Captain Loche's Catalogue as inhabiting the three provinces of Algeria, and is mentioned by Mr. Salvin as one of the most striking species in Northern Africa. In Europe it is found along the shores of the Mediterranean, from Spain to Greece. It occurs especially in Sicily, Sardinia, Corsica, Tuscany, Dalmatia, and the southern parts of France and Spain. According to Nordmann, it is found in Bessarabia; and, according to Temminck, it is very common in the States of the Church at Castello, Palo, and Civita-Vecchia, where it breeds. Savi states that along the shores of the Mediterranean it is not found further inland than from eight to nine geographical miles.

Count Mühle informs us that it does not seem to prefer the neighbourhood of water, but stops preferably in low woods or sloping hills, where the ground is dry. It is also found in gardens, especially those which are enclosed with cactus hedges, in which it builds, and upon the fruit of which it feeds. Though living exclusively in the south, it appears capable of bearing cold well, as Count Mühle frequently observed it at Christmas apparently quite lively in the myrtle and whitethorn bushes.

Of its habits and nidification, Count Mühle further observes, that it has not only perhaps the greatest share of vivacity among the members of its family, but it is by far the most numerous, and hence its manners and habits are best known. It is a restless and very lively bird, and hops continually through the low bushes, or flies from the under branches of low trees to a shrub, in pursuit of flies. It is not in the least shy of the neighbourhood of men, and may be observed all the year in gardens where people are constantly working. In the spring the male sings, while sitting on the outstretched twig of a bush, a feeble and not very melodious love-song; after ending which it creeps quickly back into the bush. The female is seldom seen, and consequently much less known than the male; its call is sharp, and similar to that of the Wagtails, or, according to Malherbe, during pairing time, like that of the *Cicadæ.*

The Sardinian Warbler builds twice or even three times in the year, in a bush or low-hanging bough, not far from the ground, say from one to three feet. The nest is tolerably compact, and is ingeniously built of blades of grass, leaves, with soft woolly plant stems, spiders' webs, and cotton woven together; the inside lined with soft small straws and horse-hair. It lays four or five eggs in the first, and only three in the second brood. The egg is greenish grey, tolerably thickly marked with small spots, darker, and forming a wreath round the larger end.

I have been favoured by my friend, Mr. Edward J. Tuck, of Wallington Rectory, Baldock, Herts., with the following account of this bird, as observed by him in France:—"*Sylvia melanocephala* was resident about Cannes, at least I saw it constantly from December to May. It was most common in the pine woods on the hills, which are very dry, and contain underwood of broom, juniper, etc. I used to see it also in the gardens nearer the sea, where it fed on the berries of the arbutus, which ripen well in that country, the shrub itself growing wild. I first heard its song on the 27th. of January, but it did not get into full song till March. Its manners are just like those of our Whitethroat in this respect, as the male will sing from a bush, and then dart off in a jerking flight into the air, still singing. They have also a rather harsh note, like the chut-chut of the Blackcap, but louder and repeated more frequently. I did not find a nest till the 22nd. of April, when I saw one in a juniper bush, among some pines on broken ground, and much exposed. It contained three young birds and a rotten egg. The nest is much like that of our Common Whitethroat, being made of straw and dried bents of grass, with a few roots and pine twigs, lined with fine grass bents. The nest measured three inches and a half across the top; depth inside two inches. The egg is nine lines long by six wide; yellowish green ground, with several darker spots all over it. We watched the pair of old birds for some time, and saw both come with food for the young; and on one occasion, on going up to the nest, the female behaved as our Partridges do,—ran along screaming on the path, with her wings fluttering as if hurt—the only instance of this in small birds I have ever seen myself, although I have heard of it. I should suppose from the occurrence of this nest so early, they have two broods in the year."

The adult male has the forehead, vertex, and back of the head as far as the nape, velvet black. This passes off gradually into the slate-blue grey of the rest of the upper parts; wings dusky black. The outer tail feathers bordered and tipped with white. The throat, belly, and cheeks bluish white, with the flanks darker.

The first primary is short; the second, third, and fourth longer and equal in size. Eyelashes brick red; the naked and swollen eyelids cinnabar red. Beak, tolerably large and strong, is black. Iris nut brown; feet blackish brown.

The female has the whole upper parts of the body brownish grey; wing and tail feathers brownish black, with a somewhat brighter edge; the white of the external tail quill is shaded with russet and grey. Throat white, and rest of under parts of the body whitish, tinted with greyish brown; the belly still darker. Eyelids and eyelashes yellowish red.

Degland remarks that specimens he received from M. A. Malherbe, killed in the neighbourhood of Gênes, were smaller than those received from the south of France. The colours were more pure, approaching to blue on the back, sides, and upon the silvery white in front of the neck and middle of the chest and belly.

My figures of this bird and its egg are from specimens sent me by M. E. Verreaux.

It is also figured by Roux, Ornith. Prov., pl. 21; Bouteil, Ornith. du Dauph., pl. 24; Gould, in B. of E., pl. 129, as Sardinian Warbler.

INSECTIVORÆ.
Family *SYLVIADÆ.*
Genus SYLVIA. *(Latham.)*

SPECTACLED WARBLER.

Sylvia conspicillata.

Sylvia conspicillata,	DE LA MARMORA; Mem. della Acad. di Torino, 1819.
" "	TEMMINCK; Man., 1820.
" "	BONAPARTE; 1838.
" "	KEYSERLING & BLASIUS; 1840.
" "	SCHINZ; Eur. Faun., 1840.
" "	SCHLEGEL; Revue, 1844.
" "	DEGLAND; 1844.
" "	MÜHLE; 1856.
Curruca conspicillata,	Z. GERBE; Dict., 1848.
Fauvette à Lunettes,	OF THE FRENCH.
Brillen Strauchsanger,	OF THE GERMANS.
Sterpazzola di Sardegna,	OF SAVI.
M' zizzee,	ARABIC; (Salvin.)

Specific Characters.—The secondaries broadly fringed with red. The two outer tail feathers almost entirely white; the two next on each side only white at the point; first primary shorter than the fifth, but longer than the sixth; the second and third equal and longest.

Length of an adult male, sent me by M. E. Verreaux, nearly five inches; carpus to tip two inches and a half; tarsus eight tenths of an inch.

This beautiful little Warbler was stated by Temminck, in his first edition, to belong exclusively to Sardinia; but, although limited in its range, it has been discovered in Sicily, Spain, in the States of the Church, and by Count Mühle in Greece. It is also included in Captain Loche's Catalogue, by whom it is stated to occur in the three provinces of Algeria.

M. O. Salvin, in the "Ibis," for July, 1859, mentions its occurrence in the Salt Lake districts of the Eastern Atlas of Africa, where it is found abundantly in the low shrubs of the uncultivated portions of that region. He states that it is very shy, and skulks from bush to bush as any one approaches. Malherbe remarks that it only breeds in Sicily, migrating in the winter; but Count Mühle doubts this statement, as he has frequently seen it in winter in Greece, in company with *S. melanocephala*. This Warbler was discovered by Marmora, in Sardinia, in 1819, and named by him *S. conspicillata*, from a black "spectacle"-looking mark between the eyes of the male bird.

According to Savi, the Spectacled Warbler dwells in Italy, among the cistus bushes on the hills, but never in shady places. In Sardinia it inhabits the bush-covered hills, from four hundred to six hundred feet above the sea level. It is a very nimble bird, rivalling the Sardinian Warbler, with which it is frequently found, in this respect. Count Mühle tells us that it sings perched on an open branch, with its feathers raised, and that its song is simple, but loud and agreeable. Its call and song-note resembles much that of the other species in the family; in fact it is so closely allied that it has frequently been confounded with *S. sub-alpina*, and has even been considered by some only a southern variety of *S. cinerea*. It is however easily distinguished from the latter by its smaller size, by the lunettes over the eyes, and by the general greater distinctness and purity of the colours. I am however at a perfect loss to imagine upon what grounds it can be separated from the "Whitethroats," and formed into a distinct genus.

The Spectacled Warbler builds in March, in low bushes, about a foot from the ground. The nest is in the shape of a blunt cone, and tolerably thick and compact. It is formed of dry grass, stems, coarse plant stalks, much down of seeds, and sometimes spiders' webs, and is lined with small roots and human as well as horse-hair, (Mühle.) The outside as well as the inside is constructed with especial care. The delicate eggs are four, rarely five in number; ground colour pale greenish grey, with fine spots, greyish and greenish grey, sometimes thicker towards the base.

In the adult male in breeding season, the vertex and cheeks are ash grey; the whole upper part of the body greyish russet, more or less marked; throat white; the rest of the inferior parts red, tinged with grey, clearer on the belly; lores and eyebrows black; cheeks white; wings blackish, with the coverts broadly fringed with lively red; tail dark brown, with the two internal barbs of the external quills white; a small and sometimes a large spot of the same colour on the extremity of the last, and a small spot on the third; beak, yellow on the borders and the basal half below, the rest blackish; feet yellowish; iris brown.

The male in autumn has the head of a less pure ash-colour; neck and mantle grey, with the feathers bordered with russet; throat white; lower part of neck bluish ash; crop and flanks red; middle of stomach whitish.

The adult female has the top of the head dark ash grey, while the black "spectacle" mark over the eyes is either less distinct or wanting. Scapularies, rump, and upper tail feathers olive brown; wings brown, with the coverts broadly fringed with russet; the uppermost feathers of the secondaries bright russet, with a conspicuous black longitudinal mark in the centre; throat greyish white; rest of inferior parts of body russet, lighter in the centre; under tail coverts, white.

The young before the first moult are of a red ash-colour above, with the throat and neck ashy white; the body below is of a reddish ash-colour, with tints lighter in the middle of the abdomen; wings brown, with the coverts largely bordered with red; tail equally brown, with the quills fringed and terminated with reddish ash, and the inferior half of the one most external on each side white.

The figures of this bird and its egg are from specimens sent me by M. E. Verreaux.

Figured also by Temminck and Laug: pl. col. 6, f. 1, old male in breeding plumage. Roux, Ornith. Prov., pl. 217, female under the name of Passerinette; Gould, B. of E.

INSECTIVORÆ.
Family *SYLVIADÆ.*
Genus SYLVIA. *(Latham.)*

MARMORA'S WARBLER.

Sylvia sarda.

Sylvia sarda,		MARMORA; Mem. della Acad. di Turin, 1819.
"	"	TEMMINCK; Man., 1820.
"	"	BONAPARTE; Birds, 1838.
"	"	KEYSERLING & BLASIUS, 1840.
"	"	SCHINZ; 1840.
"	"	SCHLEGEL; Revue, 1844.
"	"	DEGLAND; 1844.
"	"	MÜHLE; 1856.
"	*sardonia,*	VIEILLOT.
Melizophilus sarda,		GERBE; Dict. 1848.
Pyropthalma sarda,		BONAPARTE; Consp. Gen. Av.
Fauvette sarde,		OF THE FRENCH.
Sardinischer Strauchsanger,		OF THE GERMANS.
Occhiocotto sardo,		OF SAVI.

Specific Characters.—Tail wedge-shaped; the outer feather, half an inch the shortest, has a fine line upon its outer edge white; the rest of the tail feathers unicolorous. General colour smoky brown.

Length of adult male sent me by M. E. Verreaux, five inches and three tenths; from carpus to tip two inches and three tenths. Tail two inches; tarsus nine tenths of an inch.

This Warbler, though supposed to occur in Africa, has at present a known range limited to Sardinia, Sicily, and the south of France. It is said by Mühle to breed near Palermo; and Thienemann says that it is always found in company with *S. provincialis*, the Dartford Warbler. It was discovered by Marmora, in 1819, and is not unlike in plumage *S. melanocephala*, and has the same naked ring round the eyes; it may, however, be distinguished by the specific characters given above; in addition to which the beak is more slender and weak; the general colour has a more smoky tinge; the throat is also ash-grey or darker, instead of white, so that the two birds can never be confounded. Of its habits we know very little, and what we do know of them differs but slightly from those of the Dartford Warbler; its call-note is said to be a sharper and rougher cry.

According to Thienemann, its nest and eggs are similar to those of the Dartford Warbler. Degland says it builds in bushes a short distance from the ground, making a deep well-constructed nest, in which it deposits from four to six eggs, of a dirty white, slightly yellowish, with spots grey and reddish, thicker about the greater end; great diameter sixteen, small twelve millemetres.

An adult male, sent me by M. E. Verreaux, has all the upper parts of an uniform smoky brown, darker about the eyes. Throat dark ash-grey; belly and flanks pinkish, mottled with grey, approaching to black; primaries dark hair brown; tail cuneiform, with the outer quills finely edged with white; colour of the under parts of the tail lighter; beak yellow, black for one third from the point; border round the eyes cinnabar red; iris nut brown; feet yellow.

The female is generally paler in colour than the male, and the throat and belly more ash-coloured.

Figured by Temminck and Laugier, pl. enl. 24, f. 2, adult male; Gould, B. of E.

INSECTIVORÆ.
Family *SYLVIADÆ*.
Genus SYLVIA. *(Latham.)*

SECTION V.—PHYLLOPSEUSTÆ. LAUBSÄNGER.
Wood Warblers, *(Mühle.)*   MUSCIVORES, *(Temminck.)*

BONELLI'S WARBLER.

Sylvia bonelli.

Sylvia bonelli,	VIEILLOT; Faun. France, pl. 96, f. 3, et Tabl. Encycl. Ornith., 1823.
" "	KEYSERLING & BLASIUS; 1840.
" "	SCHLEGEL; 1844.
" "	MÜHLE; 1856.
" *natteri,*	TEMMINCK; Man., 1820.
" "	SCHINZ; Europ. Faun., 1840.
Phyllopneuste bonelli,	BONAPARTE; 1832.
" "	Z. GERBE; Dict., 1848.
" "	DEGLAND; 1849.
Ficedula bonellii,	SCHLEGEL; 1844.
Pouillot Bonelli, Bec-fin Natterer,	OF THE FRENCH.
Berg-Laubsänger,	OF THE GERMANS.
Natterer's Sänger,	MEYER AND BREHM.
Lui Bianco,	SAVI.

Specific Characters.—Upper parts of the body brownish grey; inferior parts white; rump and upper tail coverts yellow green;

wings reach to half the length of tail; first quill longer than the sixth, and equal to the fifth; the second longest. Tail brownish grey, with fifteen dark shaded bands crossing it; feet light brown or grey.

Length of an adult male sent me by M. E. Verreaux, four inches and a half; carpus to tip two inches and a half. Tail two inches. Tarsus eight tenths of an inch. Female about same.

This pretty little Warbler, the congener of our Willow Wren and Chiff Chaff, has a wide range in the south of Europe. It is found in Spain, in the south of France being common in Provence, in Italy, Switzerland, the Tyrol, and Salzbourg. It has occurred in the Crimea, and a single specimen is stated by Glöger to have been captured in Central Germany. It is included in Captain Loche's Catalogue of the Birds of Algeria, and, as Count Mühle remarks, would probably be found farther north, were it not often mistaken for the other Willow Wrens. It passes the winter in Arabia and Egypt. It does not appear to have been found in Greece.

The term Laubsänger, given to this group by the German naturalists, refers to their similarity in colour to the foliage of large trees, which they affect much more than bushes or shrubs. Bonelli's Warbler is found generally inhabiting wood-covered hills, preferring alders, larches, and hazels, to all other trees. Like the other Willow Wrens, it is also found frequently in gardens, and I have no doubt like them feasts upon the fruit. It is also often found on the banks of rivers.

It is a very cautious bird, and easily frightened, and then is very shy; but if unmolested it will approach dwelling-houses fearlessly.

The song is described by Count Mühle as the most monotonous of all the Laubsängers. It builds its nest in more exposed situations than its congeners, and may be easily found in dry grass meadows. It is especially partial to tall thick ferns, as well as long grass.

Malherbe's statement that it builds in poplars and other thickly-foliaged trees, is, I think with good reason, denied by Count Mühle, as being quite different to the habits of other Wood Warblers.

Thienemann says the nests are woven and bound together with dry twigs, intermixed with *Acinos vulgaris*, and lined with dry grasses, moss, and oak leaves, the entrance wide open. It contains four or five eggs, the ground colour in most of which inclines to brownish; they are also a brownish grey and a reddish brown, delicately spotted, and in many specimens forming a wreath round the greater end.

Brehm, in Bädeker's magnificent work upon the Eggs of European Birds, now publishing, says of this species, "It nests in barren places, which are scattered over with stones, leaving hollows in the ground, which are overgrown with plants, by which the nest is concealed, the stalks and blades of the growing grass being interwoven with the nest. The eggs are of a white ground, with small brown-red spots and dots thickly scattered, and of light or darker colour, wreathing like that of *S. sibalatrix*."

Temminck's account differs little from this.

Degland says the "*Pouillot Bonelli* nests in the ground in the middle of the grass, or at the bottom of underwood; its nest is similar to that of *S. sibilatrix*, and it contains from four to six eggs—short, white or reddish white, with reddish brown spots very numerous, and thickly scattered, especially at the larger end. Long diameter fifteen, shorter twelve millemetres."

The adult male has the head, nape, and upper part of the back bright yellowish olive green; the wing feathers and tail are black grey; lesser wing coverts bordered with greyish yellow green; wings brown, with the primaries bordered with bright yellow green; cheeks grey; from the nostrils a clear white stripe over the eyes. The whole of the under parts of the body clear shining white, washed with greyish on the crop, and yellow on the flanks; tail of a lighter brown than the wings, the upper three fourths of the quills edged with greenish yellow; beak brownish above, under greyish yellow; feet brownish or dark grey; iris reddish brown.

The female has the white less clear below. Before the first moult the young birds are reddish ash-colour above, silky white below, with the sides of the chest, the flanks, and the under tail coverts, bright russet; cheeks and sides of neck bright reddish ash-colour; wing coverts, primaries, and tail quills edged with bright greenish yellow.—(Degland.)

My figures of this bird and its egg are from specimens sent me by M. E. Verreaux.

Figured also by Temminck and Laugier, pl. col. 24, f. 2. Roux, Orinth. du Daph., pl. 226. Gould, B. of E., pl. 131. Bouteille, Orinth. du Dauph., pl. 26, f. 7.

Another Leaf Warbler, *Sylvia borealis*, Blasius, (the *Sylvia eversmanni* of Middendorff,) has lately, on the authority of Professor Blasius, been added to the European list. (Vide Naumannia, 1858, and Ibis, No. 4, p. 459, 1859.) This interesting Asiatic species, which has been taken in Heligoland by Herr Gätke, ought to follow the last species in this work; but in the absence of authentic specimens from which to take my figure, I must unwillingly defer a notice of it for the present.

INSECTIVORÆ.
Family SYLVIADÆ.
Genus SYLVIA. *(Latham.)*

SECTION VI.—POLYGLOTTÆ, *(Mühle.)* Spötter, or Mockers.
RIVERAINS, *(Temminck.)*

OLIVE TREE WARBLER.

Sylvia olivetorum.

Sylvia olivetorum,	STRICKLAND in GOULD'S B. of E. p. 109, 1836.
" "	TEMMINCK; Manual, 1840.
" "	MÜHLE; 1856.
Calamoherpe olivetorum,	BONAPARTE'S List., 1838.
Chloropeta olivitorum,	BONAPARTE; Gen. Consp. Av.
Salicaria olivetorum, ?	KEYSERLING AND BLASIUS. 1840.
" "	SCHLEGEL; 1844.
Hippolais olivetorum,	Z. GERBE; Revue Zool., 1844-6, Dict. in Hist. Nat., 1848.
" "	DEGLAND; 1849.
Bec-fin des oliviers,	OF THE FRENCH.
Oliven Rohrsänger,	OF THE GERMANS.

Specific Characters.—Plumage above shaded with olive brown; wings in repose do not quite reach to half the tail; first primary equal to the second and fourth, third longest; beak broad and compressed sideways anteriorly, the tip having a tooth-like or notched curve.

Length six inches five lines; wings from carpus to tip three inches and two lines; tail two inches and a half; beak from gape nine lines and three quarters, breadth of it at angle four lines; middle toe five lines, claw of the same two lines and a half; hind toe three lines and a quarter, claw of the same three lines.—SCHLEGEL.

THIS elegant species was discovered by our lamented countryman Strickland, in the Island of Zante, in 1836, and described and figured in Gould's "Birds of Europe."

Count Mühle, in his monograph of the European *Sylviadæ*, has placed this bird, *S. elaica*, and *S. hypolais*, (Latham's Pettychaps,) in a distinct section, namely, that of *Polyglottæ*, or Mockers, in consequence of the notes of some other birds being discovered in their song. Temminck placed *S. olivetorum* among his Riverains, and Keyserling and Blasius, and Schlegel, among the *Salicaria*. But Count Mühle remarks that it must be without hesitation placed near *S. hypolais*, the *Hippolais polyglottæ* of Selys-Longchamps, Gerbe, and Degland, (Latham's Pettychaps,) and consequently included in this group, forming, with *S. icterina*, the genus *Hippolais* of Gerbe.

Sylvia olivetorum has at present only one European locality, that of the olive plantations of Greece, where, however, it appears to be by no means rare. In the second volume of Naumannia, part I, page 77, it is included in a list of birds observed in the neighbourhood of Tangiers, and is said to breed there. According to Lindermayer it appears in Greece at the end of April and the beginning of May, and leaves again early in August. It has been conjectured by Baldamus, in Naumannia, part II for 1853, page 166, that this bird, though so long unknown, does actually exist in the whole of the south of Europe. But Count Mühle remarks upon this statement that if true, it must have been discovered, as no country has been so well investigated by distinguished ornithologists as Italy.

The Olive Tree Warbler is described by Count Mühle as a brisk lively bird, teasing and pecking its companions with as much pertinacity as its congener, *S. hypolais*. It lives only in olive plantations, is very shy and difficult, I am happy to say, to kill. Hence specimens are fortunately scarce, and I have been content to give a copy of Count Mühle's excellent figure. When its song betrays its whereabouts, it can hardly be distinguished from the shining olive trees, being like them, above greenish grey, below whitish—a beautiful adaptation, and one frequently met with, by which nature seems to protect her living creatures from the ruthless and often cruel and wanton destroyer.

It moves constantly about among the trees, and carefully shuns water-reeds and bushes.

It builds in the olive trees, and fastens its nest to a branch, having a twig going through it. The nest is pretty substantial, being formed of grass and lined with thistle-down, which makes it warmer, while it is even more elegant than that of most other Warblers.

The eggs are three or four in number, long oval, grey, with a rosy reddish gloss, which wears out when the egg is kept, and dark brown large angular spots, between which are scattered small dots. Count Mühle supposes they only breed once in the year.

Brehm, in Bädeker's work on European eggs, gives a description in all essential particulars the same as that above. He says "The nest is built with dry grass blades and panicles, with down of plants woven throughout, and lined with spiders' webs, small rootlets, and horse-hair, or thistle-down. The nest is exposed to view on a small branch of the olive tree; it contains four eggs, laid in the end of May or beginning of June, which are of a dull rosy red, on a grey ground, with black grey spots, and blackish dots and scrolls."

I take the description from Count Mühle.

"This bird is, when seen flying, in shape and plumage like *Sylvia orphea*, or *nisoria*, in their spring dress, from both of which, however, it is distinguished by its strong beak, broad at the root, having the upper mandible horn-colour, and the lower orange yellow. The head, and the whole of the upper parts of the body are grey, tinted with olive, which especially predominates on the upper tail coverts. A distinct conspicuous streak of greyish white from the nostrils to the eyes. The chin, root of lower mandible, and especially the upper mandible, between the nostrils and the angle of the mouth, clothed with bristly hairs. The wings grey black, with a greyish gloss or polish on them; the lesser coverts have broad greyish borders.

The other feathers of the wings edged with white, which forms a large border on the primaries and great posterior coverts. The tail is slate grey, underneath paler; the first quill bordered with white, the second with a white spot at the end, and white edge on the inner barb; the third and fourth have only a small whitish spot at the tip. The whole under part of the body is white, with a yellowish tint, the sides greyish; under tail coverts whitish; eyebrows white; iris, nut-brown; feet lead grey, with horn-coloured claws. Tarsi furnished in front with nine very distinct plates, of which the third, fourth, and fifth are largest, though the length of each does not exceed two lines."—Schlegel.

The young before the first moult, which occurs after they leave Europe, are strikingly like the young of *S. nisoria*, and *S. orphea*, but they may be readily distinguished—in addition to the already noticed characters of the formation of the beak—from *S. nisoria*, by the uniform olive grey tint of the upper part of the body, which in the latter bird is clear grey, and on the back slate-coloured. The second and third primaries of *S. nisoria* are also of almost equal length, but in *S. olivetorum* the third is considerably longer than the second. From *S. orphea* they may be distinguished by the under tail coverts and belly in *S. orphea* being tinted with rust yellow, while in the young of *S. olivetorum* the tint is greyish.

This bird has been figured by Gould, B. of E., and by Count Mühle, in his Monographie der Europäischen Sylvien.

INSECTIVORÆ.
Family SYLVIADÆ.
Genus SYLVIA. (Latham.)

OLIVACEOUS WARBLER.

Sylvia elaica.

Salicaria elaeica,	LINDERMAYER; Isis, 1843, No. 5, p. 242; Revue Zool., 1843, p. 210.
Hippolais elaeica.	Z. GERBE; Revue Zool., 1844, p. 440, 1846, p. 434; Dict. d' Hist. Nat., 1848, tome xi., p. 237.
" "	DEGLAND, 1849.
Ficedula ambigua,	SCHLEGEL; Revue, 1844.
Sylvia elaica,	MÜHLE; 1856.
Bec-fin Ambigu,	OF THE FRENCH.
Oelbaum Spötter, or Zweidentiger Laubsänger,	OF THE GERMANS.

Specific Characters —Plumage above shaded with greyish; wings in repose reach scarcely to middle of tail; first primary short, longer than the upper coverts, second and fifth equal, third and fourth longest, and equal.

Dimensions of two specimens sent me by M. E. Verreaux.—One from Greece:—from tip of beak to end of tail, five inches; length of beak seven tenths of an inch; breadth at base three tenths of an inch; wing, from carpus to tip, two inches and a half; tarsus four fifths of an inch; tail two inches. One from Algeria:—From

tip of beak to end of tail, five inches and a half; wing from carpus to tip, two inches and seven tenths; tarsus nine tenths of an inch; tail two inches and three tenths.

This is the second Warbler peculiar to the olive groves of classic Greece, that has been made known to science within the last twenty or twenty-five years. Strickland's discovery of the bird last described in 1836 led to that of the present species, by Dr. Lindermayer, in 1843. Count Mühle remarks that from these events we may conclude that many birds, especially such as from their trifling size and colouring of the plumage so readily escape observation, remain yet to be discovered. During his residence in Greece, though anxiously examining the birds of that country, this species escaped his observation.

The Olivaceous Warbler is at first sight very like the Olive Tree Warbler; but if we examine the dimensions which afford us such valuable means of differential diagnosis, we shall find a full inch difference in size. Its plumage is also altogether more of a yellow tint than that of its congener. It inhabits, however, the same localities, the olive plantations of Attica, where it also arrives late and departs early.

Count Mühle informs us, that, like the Olive Tree Warbler, it takes up its abode in the tops of the olive trees, and is never seen elsewhere. As volatile and restless as its congener, it is yet still more shy, flying, as if in terror, from one tree to another; but after a certain time returns quickly to its first resting-place, and thus it deludes those who design its capture, or the discovery of its nest.

Its piercing shrill cry may be heard repeatedly from its retreat; and it is continually engaged, with restless eagerness, in driving away every rival or feathered intruder from the sanctuary of its abode. Its song is not so unmelodious as that of the *Sylvia olivetorum*, and is more like that of other members of its family, as *S. hippolais*, the Melodious Willow Wren, or Latham's Pettychaps, of British naturalists. It is also, like the Olive Tree Warbler, very difficult to capture, keeping constantly at the top of the olive trees, hopping and gliding among the foliage, which has the same colour as itself.

The Olivaceous Warbler builds in the middle of May, in the same situations, and a similar nest to the Olive Tree Warbler. The nest is, however, smaller and less industriously made, though the materials are the same. It lays four or five eggs, pale grey green, without any shining glossy flush, covered with large black or small greenish black spots.

Brehm, in Bädeker's work, remarks of this species:—"It is an inhabitant of Greece, smaller than the Olive Tree Warbler, and of a duller plumage. It builds its nest of strips of inside bark, and fibres of roots, with thistle down, and lines it with spiders' web. It lays, beginning of June, five eggs, smaller, and duller in colour than those of the Olive Tree Warbler. Ground colour grey white, scarcely at all tinted with reddish, and marked with violet spots, and blackish and brownish points and small dots, sometimes only at the base, but at other times scattered over the whole egg."

The adult male and female have the head and all the upper parts of the body pale greyish brown, with an olive tint, more indistinct on the lower part of the back; a yellowish streak from the nostrils over the eyelids; on the angle of the mouth and chin some blackish hairs. Wings and tail greyish olive brown;

IVA HOU   A   B .
ACK-THROAT   A BL

the third, fourth, and fifth primaries have the outer web compressed or narrowed near the tip. All the wing feathers finely bordered with greyish. Tail, narrow, and notched centrally, is greyish brown; the outer quill edged with white, the second and third only on the inner web. The whole under part of the body is whitish yellow, darker in front of the chest; flanks pale yellow, tinged with olive brown; under tail coverts white. Beak above horn-colour, below pale orange yellow; feet and claws dark brown.

The bird and egg figured are from European specimens sent me by M. E. Verreaux.

Figured by Count Mühle, in his Monographie der Europäischen Sylvien, bird, nest, and eggs, the original of which were taken by Dr. Lindermayer, and are now in the museum at Ratisbon.

The next bird, which, in the natural order would be described here, is *Sylvia hippolais* of Latham and continental writers; *Hippolais polyglotta* of Selys-Longchamps, Gerbe, and Degland; Latham's Pettychaps, or Melodious Willow Wren of British writers. A single specimen however of this bird is recorded in the "Zoologist," 2228-9, as having been captured by Dr. F. Plomley, at Eythorne, near Dover, on the 15th. of June, 1848; and it is figured and described as a British bird by Mr. Morris, in the third volume of his "History of British Birds." Yarrell, Pennant, Selby, Jenyns, and Gould unfortunately adopted the specific name of *"hippolais"* in describing the Chiff-Chaff, which has caused much confusion in the synonomy of these birds.

INSECTIVORÆ.
Family SYLVIADÆ.
Genus SYLVIA. (Latham.)

VIEILLOT'S WILLOW WARBLER.

Sylvia icterina.

Sylvia icterina,	VIEILLOT; Nouveau Dict. d'Hist. Nat., tome xi., 1817, Faun. France, p. 211.
" "	TEMMINCK; Man., 2nd. Edition, 1835, p. 150.
" "	DEGLAND; Mem. de la Soc. des Sc. de Lille, 1840.
" *hippolais,*	TEMMINCK; Man., 1820, p. 222, tome i.
" "	CH. BONAPARTE; Fauna Italica, pl. 28, f. 2.
Ficedula icterina,	KEYSERLING AND BLASIUS; Die Wirbelt: 1840, p. 56.?
Hippolais polyglotta,	DE SELYS-LONGCHAMPS; Faune Belge, 1842.
" *icterina,*	Z. GERBE; Rev. Zool., 1844, p. 440, et 1846, p. 433, et Dict. d'Hist. Nat., 1848, tome xi., p. 237.
" "	DES MURS; Iconographique Ornith., Livr x., 1847.
" "	SELYS-LONGCHAMPS; Revue Zoolog., 1847.
" "	DEGLAND; Ornith. Europ., 1849.

Fauvette des Roseaux,   BUFFON; Pl. Enl., 581.
Bec-fin Icterine,   OF THE FRENCH.
Becca-fin Itterino,   OF THE ITALIANS.

Specific Characters.—Above, olive grey; primaries and tail quills brown, bordered with olive grey; below, lore, neck, sides, and superciliary ridge yellow. Wings in repose reach to middle of tail; first primary rather longer than the fourth, and nearly equal to (but still longer than) the third, the second longest.

Dimensions.—Mean of fifteen individuals measured by M. Z. Gerbe, reduced to English measurements:—Total length, from tip of beak to end of tail, 5.2 inches; from carpus to tip 3.01 inches. Tarsus one inch; beak from point to origin 0.44 inches; maximum depth (hauteur) of beak, 0.16 inches; maximum breadth of same, 0.24 inches.

THIS bird was first described as a new species by Vieillot, in 1817, for some time after which it seems to have undergone, in scientific works, many vicissitudes, —the usual fate of those subjects in natural history in which there exists a great family resemblance. It is in fact so closely allied to *Sylvia hippolais*, (Latham,)—a name unfortunately given by British writers to the Chiff-Chaff—that M. Temminck's designation and description of that bird is really applicable to *S. icterina*. M. Temminck also made another mistake in figuring a large specimen in autumn plumage of *Sylvia trochilus*, and describing the same in the second edition of the "Manual" as *S. icterina*. The Prince of Canino also figured this bird in the "Fauna Italica" as *Sylvia hippolais;* and the Honourable M. De Selys-Longchamps made a similar mistake in the "Faune Belge," in 1842. It is to M. Z. Gerbe that we are indebted in the "Revue Zoologique" for 1846, and in the "Dictionary of Natural History" for 1848, for restoring this bird to its proper place among the European species, and for indicating in the clearest manner, those essential points of difference which must prevent any confusion of the species—or perhaps I should rather say race—in future. In a work of singular beauty and elegance, the "Iconographique Ornithologique," M. Des Murs has collected all the evidence that could be adduced upon the subject, and has given an excellent figure of the bird, contrasting it on the same sheet with *S. hippolais*, (Latham.) I have copied M. Des Murs' figure for the following reasons:—First, specimens of the true *Sylvia icterina* are rare and difficult to meet with; and secondly, as I consider the species stands upon the testimony collected by M. Des Murs, it is only right I should give a figure of the bird indicated by him. Further testimony has been offered to the correctness of M. Gerbe's description by the distinguished naturalist, Blasius, both in "Naumannia," and in Count Mühle's "Monograph." It is better, therefore, that we should distinctly know which is the bird meant by M. Gerbe himself, rather than trust to specimens which after all may not prove to be correct species.

M. Gerbe informs us that it is very difficult to distinguish *Sylvia icterina* from *Sylvia hippolais*, the *Hippolais polyglotta* of continental writers, the Latham's Pettychaps, or Melodious Willow Wren which is figured and described as a British species in Morris's "British Birds," and Yarrell's supplement, from a specimen recorded in the "Zoologist," 2228, as having been captured in England. We have the two birds the same size, colour, and form, but they may be distinguished by the following characters.

In *Sylvia hippolais* the wing in repose does not reach half way to the tail, and the first primary is equal or nearly equal to the fifth. In *S. icterina* the wing is always longer by two fifths of an inch or more, than that of *S. hippolais*, and it consequently reaches beyond the middle of the tail, and the first primary is equal in length to the third: and these distinctions are constant both in male and female.

Sylvia icterina is found, according to M. Gerbe, in France, Belgium, Austria, Sardinia, Liguria, and probably in a great part of Italy and Sicily. It has been captured in the neighbourhood of Genes, and was found plentifully by M. Gerbe, not only at Nice but in the valleys on the shores of the Mediterranean. It has also been captured in breeding plumage by L'Abbé Caire, in the neighbourhood of Moustiers, in the Basses Alps. In the "Revue Zoologique" for 1847, M. de Selys-Longchamps says that it is very common in Belgium, particularly in Liége and Brabant; and he gives the following dates of its first appearance in that country for six years:—In 1841, May 4th.; 1842, May 12th.; 1843, May 17th.; 1844, May 15th.; 1845, May 14th.; 1846, May 11th. "It loves," continues this writer, "to occupy damp groves and willow plantations near the water. It is also very common in the dry hills planted with vines and fruit trees, in the neighbourhood of Liége. Even the smallest garden in Liége has its annual nest, and during the months of May, June, and July, the male sings constantly a varied and powerful song, somewhat like that of *Sylvia palustris*, but livelier and more gay. It also imitates the Greenfinch, the call-notes of the Chimney Swallow, the Golden Oriole, and the Woodchat Shrike. Its usual call-note is similar to that of the other Wood Warblers. It loves warmth, and without doubt passes the winter in southern countries, for it leaves in August, and those in captivity appear to suffer much from cold.

In short I have remarked we do not often find it in the mountainous districts between the Meuse and Prussia, which are doubtless too high and cold. It does not exist in England, but is found in Flanders and Artois."

M. Sundeval, Curator of the Stockholm Museum, also remarks of this bird:—"It catches insects, like the Flycatchers. It is a courageous and quarrelsome bird, perpetually in motion, except when singing, and then it chooses an elevated branch of a tree. Its nest is something like the Blackcap's in form, but contains a greater number of feathers; it is most frequently placed in lilac bushes or rather low fruit trees. The eggs are four or five, reddish lilac, with black dots thickly scattered over them."

Of the habits of *S. icterina* M. Gerbe says:—"It lives on the shaded sides of hills in fertile and humid valleys; it delights to frequent willow oziers, and, it is said, reeds. I have frequently found it in olive plantations. Its food differs but little from that of *S. polyglotta*, like which, it catches insects on the wing. I frequently found debris of clytræ in the gizzard, mixed with snails. It often adds to its regime fruits and berries. It lays four or five eggs, slightly larger than those of *S. polyglotta*, but having the same form and distribution of colours."

The male and female have the upper part of the head and neck olive grey; forehead olive, saturated with yellow; rump bright greenish ash. Inferior parts, space between beak and eyes, superciliary ridge, cheeks, and sides of neck, yellow; flanks grey brown, shaded with yellow; wings brown, the primaries being bordered with greenish grey; secondaries broadly fringed with yellowish white, and near their origin with greenish; greater wing coverts have a green ash-colour on the edges; tail brown, bordered with yellowish grey, the external quills darker, the others finely bordered with russet white; iris, dark brown; superior mandible brown, inferior yellow; tarsi and feet bluish; claws brown.

The young before and after their first moult are not very different from the adult birds; all the colours are paler, and the fringes of the wings and tail, instead of being whitish or greenish, are of a bright yellowish green.

Figured by Vieillot, in the Faun. Fr., pl. 96, figs. 2 and 3; by Buffon, Pl. Enl. 581, as *Fauvette des Roseaux;* Prince Charles Bonaparte, in Fauna Italica, pl. 28, fig. 2, under the name of *Sylvia hippolais;* and by M. des Murs, Iconographic Ornithologique, livraison x., 1847, from which my figure is taken.

INSECTIVORÆ.
Family SYLVIADÆ.
Genus SYLVIA. (Latham.)

BLACK-THROATED WARBLER.

Sylvia virens.

Sylvia virens,	LATHAM. WILSON.
" "	AUDUBON; Plate 399.
" "	GATKE; in Naumannia, 1859, p. 425.
Sylvicola virens,	SWAINSON. BONAPARTE; Birds of E. and A., p. 22.

Specific Characters.—Plumage above yellowish olive green; throat black, the feathers in winter being edged with yellowish white; first and fourth primary of equal length, the second and third a trifle longer. Length five inches.

THE lonely rocky island which rears up its bold red-looking front to the eyes of the traveller, as he steams near the mouth of the Elbe, on his way to Hamburg, seems destined to be a rich field of discovery to the European ornithologist. So many new forms have lately turned up in Heligoland, that the celebrated Professor of Brunswick—Blasius, has been induced to make an excursion thither, and examine for himself the treasures which have been collected

by the resident naturalist, Herr Gätke. The results of this visit, as well as that of Herr Gätke's experience, have been communicated to the ornithological world in the last number of "Naumannia."

And now that we have seen the original articles, I will take the liberty of correcting a slight error which I made in the note at page 48. *Phyllopneuste borealis* is, I find, claimed by Blasius as an entirely new species, and he gives diagnostic descriptions of this bird, as compared with those of *P. javonica* and *P. icterina*, between which two species it appears to be intermediate. I shall offer a few remarks upon this and other *Sylviadæ*, when I come to the end of this interesting group. With regard to the general result of the observations of Blasius and Herr Gätke in Heligoland, they are extremely interesting.

Blasius gives at least four species entirely new to Europe, of which two are American forms; while Herr Gätke mentions no less than twenty-three species which he considers new to Europe, observed in the island up to 1858; and he concludes his paper with the very natural exclamation, what will 1859 bring forth?

The most interesting part of these papers is the number of American species which the lists contain: among them is the subject of the present notice. Perhaps the evidence adduced by these lists may rather remove the hasty scepticism with which my introduction of the Bald Eagle into the European fauna was met.

Sylvia virens is an inhabitant of America generally. The specimen which Herr Gätke records was killed on Heligoland, on the 19th. of October, 1858, by a small boy with a pea-shooter; and in order that there may be no mistake about the determination of the species, I will here give a verbatim translation of Herr Gätke's very clear description. The bird is in its winter plumage, and thus differs from my figure, which is in its summer dress.

"The upper part of the head, back, and rump, a beautiful clear yellowish olive green, more inclining to yellow towards the rump; forehead, a broad stripe over the eyes, and sides of the throat, very beautiful clear yellow; from the beak to the eyes is a blackish stripe, which is continued on the other side of the eye, and terminates, or is blended with the colouring of the ear coverts. Chin, and front of head and neck, are clear black, the feathers having yellowish white edges, which from the chin very much conceal the ground colour. According to Wilson these bright borders wholly disappear in summer, and leave these parts clear and shining black, (see figure,) which has given to the bird its name of "Black-throated Warbler." Sides of breast, belly, and under tail coverts are yellowish white, having on each side two broad black stripes.

Wing and tail feathers are black, with bluish ash grey borders, which on the back of the wing become almost white; greater wing coverts have broad white tips, the smaller entirely white, by which two shining white bands across the wing are formed. Both the outer tail feathers are almost entirely white, having on the outer web only a faint black stripe, which is broader at the tip, and towards the rump is almost lost in the black shaft; the third tail feathers also have on the inner webs a large white spot. Beak, feet, and iris, brown black.

The form of this little bird is elegant, though the beak is rather strong in proportion to the size of the bird; the feet are somewhat feeble. The tail is almost evenly notched; the outer pair of feathers are rather the shortest. Wings are tolerably pointed; the first and fourth primaries are of equal length, the second and third only a trifle longer; the second, third, and fourth are near the tips of the outer web very strongly notched.

The whole length of the bird, from the forehead to the point of the tail, five inches; length of the wing in repose two inches and six lines; of the tail two inches; an inch of the tail uncovered by the wings. Beak five lines; tarsus nine lines; middle toe three lines; claws two lines; hinder toe three lines, and the very curved claw of ditto two lines and a half."

The bird has been preserved in spirits.

Sylvia virens belongs to the group of Leaf Warblers, but as its existence in Europe was not known by me sufficiently early to place it in that section, I have introduced it here, rather than defer a notice of it to the end of the work. Its habits are so well described by Audubon, that I take the liberty of transcribing the following from his large octavo work upon the "Birds of America."

"I have traced this species from Texas to Newfoundland, although at considerable intervals, along our Atlantic coasts, it being of rare occurrence, or wanting in some parts, while in others it is abundant: but in no portion of the United States have I met with it so plentiful as around Eastport, in Maine, where I saw it in the month of May. Many remain all summer in that State, as well as in Massachusetts and the northern parts of New York; and some are found at that season even in the higher portions of Pennsylvania. On the coast of Labrador it was not observed by me or any of my party; and it is not mentioned by Dr. Richardson, as having been seen in the Fur Countries. Its habits are intermediate between those of many of our Warblers and the Vireos, the notes of which latter it in a great measure assumes.

It usually makes its appearance in Maryland and New Jersey about the first week in May, when it is observed to be actively engaged in searching for food, regardless, as it were, of the presence of man. Its movements when proceeding northward are rapid, and it advances through the woods solitarily, or nearly so, it being seldom that more than two or three are found together at this time, or indeed during the breeding-season, at which period each pair appropriates to itself a certain extent of ground. Its retrograde march is also rapid, and by the middle of October they all seem to have passed beyond the limits of our most southern States.

The food of this species consists during the summer months of various kinds of flies and caterpillars, many of the former of which it captures by darting after them from its perch, in the manner of Flycatchers and Vireos, emitting, like them also, a clicking sound from its bill. In the autumn it is often seen feeding on small berries of various sorts, in which respect also it resembles the birds just mentioned.

I never found the nest of this bird, of which, however, Mr. Nuttall has given a minute description, which I shall here, with his permission, place before you:—'Last summer, 1830, on the 8th. of June, I was so fortunate as to find a nest of this species, in a perfectly solitary situation, on the Blue Hills of Milton. The female was sitting, and about to hatch. The nest was in a low, thick, and stunted Virginia juniper. When I approached near to the nest the female stood motionless on its edge, and peeped down in such a manner that I imagined her to be a young bird; she then darted directly to the earth and ran, but when deceived I sought her on the ground she had very expertly disappeared, and I now found the nest to contain four roundish eggs, white, inclining to flesh-colour, variegated, more particularly at the great end, with pale purplish points of various sizes, interspersed with other large spots of brown and blackish. The nest was formed of circularly entwined fine stripes of the inner bark of the juniper, and the tough white fibrous bark of some other plant, bedded with soft feathers of the Robin, and lined with a few horse-hairs and some slender tops of bentgrass, (*Agrostis*.')

My friend describes the notes of this species as follows:—'This simple, rather drawling, and somewhat plaintive song, uttered at short intervals, resembles the syllables 'te dé territica,' sometimes 'tederisca,' pronounced pretty loud and slow, and the tones proceeding from high to low.' These notes I am well acquainted with, but none can describe the songs of our different species like Nuttall."

My figure is taken from Audubon's large folio work, the drawing being kindly executed for me by Mr. Sinclair, of Glasgow, to whom I shall also on another occasion have to express my obligations.

Figured also by Wilson and Nuttall.

INSECTIVORÆ.
Family *SYLVIADÆ.*
Genus SYLVIA. *(Latham.)*

SECTION VII.—CALAMODYTÆ. Rohrsänger, Reed Warblers, (*Mühle.*) RIVERAINS, (*Temminck.*)

MARSH WARBLER.

Sylvia palustris.

Sylvia palustris,	BECHSTEIN; Nat. Deut., 1801.
" "	MEYER AND WOLFF; Tach. der Deuts., 1810.
" "	NAUMANN; Natur. der Vogel Deut.
" "	TEMMINCK; Man., 1820.
" "	VIEILLOT; Faun. Fran.
" "	SCHINZ; Europ. Faun., 1840.
" "	MÜHLE, 1856.
" *stripera,*	VIEILLOT; Dict., 1817, tome xi., p. 182. ?
Calamoherpe palustris,	BOIE; Isis, 1826, p. 972.
" "	BONAPARTE; Birds, 1838.
" "	Z. GERBE; Dict., 1848.
" "	DEGLAND; 1849.
Salicaria palustris,	KEYSERLING AND BLASIUS; Die Wirbelt, 1840.
" "	SCHLEGEL; Revue, 1844.
Riverain Verderolle,	OF THE FRENCH.
Sumpf-Rohrsänger,	OF THE GERMANS.

Specific Characters.—Upper parts of the plumage shaded with greenish or olive grey; a white streak over the eyes, and the under parts of the body also white, with an ochreous tint. Tail coverts yellowish, edged with olive grey; both the outer tail quills pale whitish at the end; the longest wing primary longer than the longest secondary. Length of an adult male sent me by M. E. Verreaux, five inches and a half; carpus to tip, two inches and a quarter; tail two inches and a quarter; tarsus nine tenths of an inch.

We have now arrived at Count Mühle's last section of the *Sylviadæ*, containing the interesting and large group of Reed Warblers. The Sedge and Reed Warblers of our own island are so well known that it is hardly necessary to say anything about the habits of the family, which are very similar in all the species. They may, however, be distinguished by the following characters. The forehead is narrow and flat, the feet are strong, and the claws long and slender; wings short, and the tail round or cuneiform. The distinctions of sex are not strongly marked, and there is generally a bright whitish or cream-coloured stripe extending over the eyes. The young moult in the spring for the first time. They are found principally in northern climates, and generally near water or marshes, and are not found among high trees or in mountainous districts. They arrive late and migrate early, and there is a great similarity in their song. They are insectivorous, and build in reeds or bushes, having always a stalk or branch passing through the nest, which is so formed and fixed, that although swayed about by the wind so as nearly to touch the water, the eggs do not fall out.

The Marsh Warbler, the first which I am called upon to notice, is very similar in external form and colour to our Reed Warbler, but differs considerably from it in its song and nidification. It has a wide range in Europe, being found in Russia, Germany, Holland, Belgium, Switzerland, Italy, and France. It does not seem to go farther north than Denmark. Count Mühle states that it is found in the whole of North and South Africa, and in the south-west of Asia. I cannot, however, find it in either Hodgson's Catalogue, or that of Mr. A. Leith Adams, of the birds of India, published in the "Zoological Transactions" for November, 1858, and May, 1859—two exceedingly interesting and valuable contributions to Indian ornithology; neither is it in Mr. Salvin's list of the "Birds of the Eastern Atlas of Africa," or in Captain Loche's "Catalogue of the Birds of Algeria;" but there is no doubt it may have been confounded in the above lists with *Sylvia arundinacea*.

In Europe it is found, according to Temminck, plentifully on the banks of the Po and the Danube; and Degland records its appearance in the department of Nord. A male was killed in 1843, in the neighbourhood of Bergnes, and subsequently every year others at the same place. M. Baillon has procured it from Abbeville, and M. Gerbe plentifully from the Basses Alps. It is generally distributed in Germany, appearing in May, and leaving again in September. It is found, not in thick reed and sedge clumps, but chiefly on the banks of rivers, where the brushwood is low and mixed with reeds, high grass, sedges, etc., closely grown together.

The following is from Count Mühle's description of its habits:—"The Marsh Warbler is a very neat merry bird. Quick in all its movements, it is equally active in skipping through the bushes as in flight. Bold and enterprising, it becomes also arrogant and tyrannical in its combats with other birds dwelling around it. It seems never to repose, and hardly does the eye catch it than its voice is heard perhaps a hundred paces farther off. Of all the Reed Warblers it has the most beautiful and varied song, enlivening an otherwise dull and monotonous part of the country. It is a master in imitation, and knows quite well how to blend, in a delightful whole, the different songs of the surrounding birds. In warm summer it sings all night through, and so charmingly in the stillness of the time and scene, that we are tempted to compare it with the Nightingale. Its call-note is not often heard, but is similar to that of other Reed Warblers. Its nest is never placed over water, nor even over marshy ground; it is found in shrubs and bushes from one to three feet above the ground: the inside is deep, like that of other Reed Warblers' nests, and formed of delicate grass blades, straws, nettle fibres, and spiders' webs. It is lined with very fine straw and a tolerable quantity of horse-hair. It lays four or five eggs, which are bluish white, sparingly spotted with delicate grey dots, and olive brown and ash grey spots."

Brehm, in Bädeker's work upon European eggs, says of this bird:—"It builds in bushes in meadows and on the banks of ditches, rivers, ponds, and lakes. The nest is made of dry grass and straws, with panicles, and interwoven with strips of inner bark and horse-hair outside. The rim is only very slightly drawn in. It has a loose substructure, and is by this and its half-globular form, suspended on dry ground between the branches of the bushes or nettles, easily distinguished from the strongly-formed nest of *S. arundinacea*, which is moreover built over water. It lays five or six eggs the beginning of June, which have a bluish white ground, with pale violet and clear brown spots in the texture of the shell, and delicate dark brown spots on the surface, mingled with which are a number of black dots. The ground colour also in many fresh eggs is green, but clear and very different from the muddy tint of the egg of the Reed Warbler. The female sits daily for some hours, but the male takes his turn. Incubation lasts thirteen days."

I have been thus particular in quoting the habits and nidification of this bird, as they are the principal means by which the species, though undoubted, is distinguished from the Reed Warbler. Its powers of imitation are indeed remarkable. M. L'Abbe Caire writes to M. Gerbe:—"This species sings most admirably, imitating with exactitude the notes of the Goldfinch, the Chaffinch, and the Blackbird, as well as all the other birds which frequent its neighbourhood. Its song is richer in variation than that of the Nightingale, and it can be listened to from morning to night."

I think it very probable that this bird is an inhabitant of Great Britain, though hitherto confounded with the Reed Warbler. I think I have myself taken the nest; and Mr. Sweet's bird, mentioned by Mr. Yarrell, was probably this species.

The male and female in breeding plumage are greenish olive grey, the rump somewhat paler green; inferior parts of a white russet, lightest on the chin and throat, having a yellow tint on the sides of the neck and belly; the lores, and a line above the eyes, reddish white. Wings brown, bordered with ash; tail same, bordered with greyish; first primary very short, second a little shorter than the third, which is the longest, and which is twenty millemetres longer than the longest secondary, that of the Reed Warbler being only sixteen. Beak above, black brown, lower mandible yellow flesh-

colour; length five lines and a half, and two lines at the base, being shorter and broader than that of the Reed Warbler; the gape, which is orange yellow, is garnished with three or four strong black hairs; iris dark brown; feet yellow flesh-colour; claws darker.

The young birds are above clear olive grey, and underneath slightly inclining to a rusty yellow.

My figures of this bird and its egg are from specimens sent me by M. E. Verreaux.

It is figured also by Naumann, in his Naturgeschichte Der Vögel Deutschlands, Taf. 81, fig. 3, (male;) and by Gould, B. of E. The egg is also figured in Bädeker's work.

INSECTIVORÆ.
Family SYLVIADÆ.
Genus Sylvia. *(Latham.)*

BOOTED REED WARBLER.

Sylvia scita.

Motacilla salicaria,	Pallas; Zoograph. ross. Asiat., i., p. 492.
Sylvia caligata,	Lichtenstein; in Eversmann's "Reise nach Buchara," p. 128.
" *scita,*	Eversmann; Addend., iii., page 12-13; and in Bullet. de la Societe Imp. de Moscow, 1848, No. 1, p. 225.
" "	Mühle; Monograph der Europ. Sylvien, 1856.
" "	Thienemann; Fortpflanzung, v., 199.
Lusciola caligata,	Keyserling and Blasius, 1840.
Salicaria caligata,	Schlegel; Revue, 1844.
Calamorphe caligata,	Degland; 1849.
Riverain botté,	Of the French.
Gestiefelter Rohrsänger, and Schmucker Rohrsänger,	Of the Germans.

Specific Characters.—The tarsi in front furnished with four scales, occupying two thirds of the whole length; the first and last

small, the third double the length of the first. All the tail feathers rounded, the middle one somewhat shortened, and edged with pale grey brown. Length four inches and two fifths; wings two inches and four lines; tail two inches; tarsi nine lines and a half; middle toe five lines.

THERE has been much written about this little bird, whose history appears to be as follows.—Pallas, in his "Zoography of Asiatic Russia," describes a small Reed Warbler, under the designation of *Motacilla salicaria;* the Warblers in those days being mixed up with the Wagtails. In the history of Eversmann's travels to Bucharest, Lichtenstein, the German naturalist, has noticed a bird, now in the Berlin Museum, labelled "*Sylvia caligata*, Siberia, Eversmann," in the following words:—"A new species, and distinct from all our European Reed Warblers, which Pallas, under the mistaken name of *Motacilla salicaria*, very fully and correctly described."

It resembles *Sylvia arundinacea*, Latham, in its youthful plumage, but it may be distinguished as follows: —"The length from the tip of the beak to the rump is only two inches and five lines; the tail is about two inches one line; the beak is much smaller, only five lines and a half long. The tarsus is nine lines; the superciliary streak not clearly developed, and it is booted to the root of the toes with scales. The construction of the wing is also different: the second primary is of the same length as the sixth, and the third, fourth, and fifth are the longest, whilst in *arundinacea* the fifth is shorter than the second; also the fourth, fifth, and sixth are contracted in the outer web. The legs are of a bright colour, and the first year's plumage of *arundinacea* is much paler."

Keyserling and Blasius also describe the Berlin specimen minutely, and consider it synonymous with Pallas's *Motacilla salicaria*, which view is also taken by Schlegel in an elaborate analysis in his "Revue Critique." Lastly, Count Mühle, after careful examination of the specimen in the Berlin Museum, identifies it with a specimen he had killed in Greece. Eversmann, having in 1842-3 published an addenda to Pallas's "Zoography," described the bird which he had discovered as *Sylvia scita*. Thus, though the identity of *Motacilla salicaria* and *S. scita* may be still open to doubt, and is in fact doubted by Count Mühle, it is quite certain that the latter bird, captured in Greece, and described and figured in his work, is identical with the *S. scita* of Eversmann, thus establishing clearly its title to the distinction of an European species.

The Booted Warbler has only been found in Siberia, Russia, and Greece. Eversmann found it on the banks of the rivers in the Ural Mountains. It is described by Pallas as inhabiting the banks of rivers, among the willows. It hangs on the stems of the trees, and is continually in motion, and singing most agreeably. It constructs in the forks of the branches a nest composed of grass, and it lays four or five eggs.

Thienemann figures the egg from specimens sent from the Volga, but I think this source too doubtful for reliance. Altogether we want a great deal more information about this species.

The upper parts are of a pale and dirty olive-colour; the inferior whitish, but the throat is of a pure white. Primaries and tail brownish grey; middle tail feathers with lighter edges, the external ones edged with whitish on both sides: the following are only edged with this colour on the inner barbs and at the tip. Beak black, approaching to white at the base; feet brown.

My figure is taken from the drawing of the original specimen of this bird by Count Mühle. The egg, though given with doubt, is from Thienemann.

INSECTIVORÆ.
Family *SYLVIADÆ.*
Genus Sylvia. *(Latham.)*

AQUATIC WARBLER.

Sylvia aquatica.

Motacilla aquatica,	Gmelin; Syst., vol. i., p. 953.
Sylvia aquatica,	Latham; Ind., 1790, vol. ii., p. 510.
" "	Temminck; Man., 1820.
" "	Schinz; Europ. Faun., 1840.
" "	Mühle, 1856.
" *salicaria,*	Meyer and Wolff, 1810.
" *paludicola,*	Vieillot; Dict., 1817.
Calamodyta schœnobænus, et Cariceti,	Bonaparte, 1838. Gerbe; Dict., 1848.
Calamodyta aquatica,	Degland, 1849.
Salicaria aquatica,	Keyserling & Blasius, 1840.
" "	Schlegel, 1844.
" "	Gould; B. of E.
Riverain aquatique,	Of the French.
Seggen Rohrsänger,	Of the Germans.

Specific Characters.—A large band of yellowish white or yellow over the eyes; on the head two large longitudinal black bands, separated by a reddish yellow band; the under tail coverts and the rump marked with oblong blackish spots; tarsi flesh-colour. Length of a male specimen sent me M. E. Verreaux, four inches and four tenths; carpus to tip two inches and a half; tarsus six tenths of an inch.

THE Aquatic Warbler has a somewhat limited range. It is only known with certainty to breed in Germany and Holland. According to Temminck it is only an accidental visitor in the latter country, but Mühle gives this as one of its breeding places. It is plentiful in Italy and the south of France during the passage. It is found in Switzerland and Sardinia, on the banks of the Var and Rhone, and in the marshes surrounding Arles. It is also found at Dieppe, and in the marshes about Lille. It occurs in Algeria, as stated by Captain Loche; and Mr. Salvin, in his "Five Months' Bird-nesting in the Eastern Atlas," in the "Ibis" for July, 1859, says, "At the head of the little marsh of Aïn Djendeli, I more than once observed a pair of this Warbler. We afterwards found it more abundant at Zana, where it was breeding. In its habits it much resembles the Common Reed Warbler, *(C. arundinacea ;)* the eggs also are similar."

"It is really plentiful nowhere," says Count Mühle, "and it dwells preferably in large wild swamps. In summer it need only be sought for where the water is cooped up almost knee-deep, with ditches and dry necks of land running into it, and covered with bushes, high grass, rushes, and reeds. In autumn it may be found in more cultivated ground."

"It is a very restless and lively bird, and also crafty and cunning. It creeps with great agility through the twigs and stalks of the thick swampy plants, in which it excels all other Reed Warblers. It may be seen gliding along near the ground, like a mouse; it never hops on the ground, but goes along step by step. On the stalks and perpendicular stems of plants it may be seen running up and down with such agility that it seems to slide along without using its feet at all. Its call is like the rest of the Reed Warblers'; its love song, though loud, is also pleasant, and comes almost always from the depth of the reed-beds, and seldom from the summit of the stalks: it is, however, proportionally often heard among trees. It builds its nest in the swamp; the exterior is formed of coarse grass tops, intertwined with delicate straws: it is lined inside with horse-hair. It is placed between the slender twigs of small bushes, and always especially found in isolated marshy places intersected with ditches. It lays in the beginning of May four or five, rarely six, eggs, grey-greenish or grey-yellowish ground, with spots more or less strongly marked, darker than the ground colour."

Brehm, in Bädeker's "European Eggs," says of this species:—"It breeds in Holland, Greece, Germany, and probably in Switzerland and Italy. At the end of April we hear its nuptial song in the marshes, among the bulrushes, reeds, and bog plants which grow there. Its nest may be found the end of May, containing five or six eggs, deep under a clump of sedges, in the grass behind rubbish, or on the bank of a hedge near water, hanging on the stalks of a plant. It is unlike the nest of the Sedge Warbler in being smaller, but is built of the same materials, namely, small rootlets, mixed with strips of reed and straw, under which is also some horse-hair. The eggs are smaller, brighter, smoother, and more shining than those of *S. phragmites*, and are often marked with hair-streaks. Very often the markings are so faint that the egg appears unicolorous. Once we found a nest containing eggs washed with carmine. The male sits but little, the female most assiduously. Incubation thirteen days."

M. Moquin-Tandon has kindly sent me the drawing from which the figure of my egg is taken, with the following remarks:—"This egg comes from the environs of Angers. I had it from M. de Baracé, a distinguished ornithologist. The nest is in the form of a cone, cleverly constructed. It contains four or five eggs, of a dirty greenish grey, with olive spots more or less dark, generally forming a wreath at the thicker end. I have seen some specimens of a deeper grey. Great diameter seventeen to eighteen millemetres, small twelve to thirteen."

The male and female in breeding-plumage have the upper parts a pretty ash grey, passing to yellow red on the rump and upper tail coverts, with black spots, forming two longitudinal bands on the vertex, small and less apparent on the neck, deep and large on the back, narrow on the tail coverts; inferior parts of a yellow russet, very clear, becoming white on the throat and on the middle of the belly; a large superciliary band, the same colour as the throat; another, brown, above the ears, larger over the auditory orifice; wing coverts brown, thickly bordered with ash grey; primaries blackish, bordered with grey; tail quills brown, bordered with greyish, the most external of each having an ashy tint. Beak brown above, yellowish below and on the edges of the mandibles; feet yellowish, with the bottom of the toes yellow; iris bright brown; first primary short, second and third equal and the longest.

In autumn the plumage is yellow russet above, with black spots in the centre of the feathers, as in spring; below of a clearer russet; all the quills of the wings and tail bordered with yellowish red or grey.

Young after the first moult resemble the old birds, but have on the neck and flanks brown striæ, more or less numerous.—(Degland.)

My figure is taken from a male in breeding-plumage sent me by M. E. Verreaux.

Figured by Naumann, Taf. 82, who also figures on the same plate the so-called species *cariceti*, which is, however, only a dark variety of *aquatica*. Gould, B. of E.

INSECTIVORÆ.
Family *SYLVIADÆ.*
Genus SYLVIA. *(Latham.)*

MOUSTACHED WARBLER.

Sylvia melanopogen.

Sylvia melanopogen,	TEMMINCK; Man., 2nd. Ed., p. 121, 1835.
" "	SCHINZ; Europ. Faun., 1840.
" "	MÜHLE; Monograph, 1856.
Calamodyta melanopogen,	BONAPARTE, 1838.
Salicaria melanopogen,	KEYSEBLING AND BLASIUS, 1840.
" "	SCHLEGEL; Revue, 1844.
Cettia melanopogen,	Z. GERBE; Dict., 1848.
" "	DEGLAND, 1849.
Bec-fin à Moustaches Noires,	OF THE FRENCH.
Schwarzbartiger Rohrsänger,	OF THE GERMANS.
Torapaglie Castagnolo,	OF SAVI.

Specific Characters.—Beak very slender, and drawn inwards; eyebrows broad and white; the vertex, lore, and a spot behind the eyes black; twelve quills in the tail; tarsi blackish. Length four inches and a half; carpus to tip two inches and one fifth; tail one inch and a half; tarsus nine tenths of an inch; beak, from gape, half an inch.

The Moustached Warbler is a rare bird, for, notwithstanding the statement of Temminck that it is common in the neighbourhood of Rome, near Ragusa, and in Tuscany, but few specimens are found in cabinets. It occurs, however, in Sicily, Italy, the south of France, Greece, in the swamps of Lentini, and at Syracuse. Temminck says that it is *"très commune"* about the lake Castiglione, and Ostia. However this may be we know very little about its habits, more than has been mentioned by M. Cantraine, who collected specimens for Temminck in Italy, and who remarks that it always lives in the swamps, and in the bushes which surround them. It climbs along the reed-stems like the rest of its family, and lets its sharp singing notes be heard clearly enough. Like *Sylvia aquatica*, it also runs along the reed-stems and water-plants, above the surface of the water. It is not shy, and frequently sits on the tops of the reeds. Those which M. Cantraine killed in the winter were all males.

M. Moquin-Tandon has kindly sent me a beautiful drawing of the egg, (which I have figured,) with the following remarks:—"From the neighbourhood of Montpellier, from whence it was sent me by M. Lebrun. This Warbler builds among the reeds. Its nest is small, in the form of a deep cup, and composed of fibrils and roots and leaves of small grasses, and the interior is lined with horse-hair and wool. It contains four or five eggs, having an azure white ground, with brown spots, larger and more thickly scattered round the larger end. Great diameter fourteen millemetres, small eleven."

A male in breeding plumage sent me by M. E. Verreaux, has the head dark black; nape, back, and rump rich nut brown, with longitudinal rays of black on the middle of the feathers of the back; throat,

crop, and upper part of the belly pure white, slightly shaded about the crop with russet; flanks, lower part of belly, and under tail coverts a lively russet; the lores and a spot behind the eyes, black; superciliary ridge white, becoming broader towards the nape. Wings and tail dark brown, the inner barbs of the primaries bordered with white; beak and legs brown; iris nut brown. First primary very short, the second shorter than the third, which is somewhat shorter than the fourth, the fourth and fifth almost equally long, the latter longest: the exterior web of the fourth and fifth contracted towards the end.

Degland says that in the autumnal plumage the upper parts are of a tint less dark, with the black lines on the centre of the feathers of the head, and the borders of those on the body redder. The white of the neck, crop, and stomach less pure; the crop and flanks of a darker reddish brown. The young before the first moult are of a lighter brown, with more olive-colour above.

Figured by Temminck and Laug, pl. col., 245; by Roux, Ornith. Prov., pl. 233; Savi, Ornith. Tosc.; Mühle, Monogr. der Eur. Sylv.

INSECTIVORÆ.
Family SYLVIADÆ.
Genus SYLVIA. *(Latham.)*

FANTAIL WARBLER.

Sylvia cisticola.

Sylvia cisticola,	TEMMINCK; Man., 228, 1820.
" "	VIEILLOT; Faun. Fr., p. 227.
" "	SAVI; Ornith. Tosc., 1827.
" "	SCHINZ; Europ. Faun., 1840.
" "	MÜHLE; Monog., 1856.
" *typus,*	RÜPPELL; Neue Wirb. Vögel, p. 113.
Cisticola schœnicola,	BONAPARTE, 1838.
" "	DURAZZO; Ucelli Liguri, 1840.
" "	Z. GERBE; Dict., 1848.
" "	DEGLAND, 1849.
Salicaria cisticola,	KEYSERLING & BLASIUS, 1840.
" "	SCHLEGEL; Revue, 1844.
Riverain Cisticole,	OF THE FRENCH.
Cisten Rohrsänger,	OF THE GERMANS.
Beccamosche,	OF SAVI.

Specific Characters.—The fan-shaped tail grey brown, each tail feather tipped with white, and with a black spot near the extremity of each; no white band over the eyes; beak curved downwards. Length of a specimen sent me by M. E. Verreaux, three inches and four fifths; from carpus to tip two inches and one fifth; tail one inch and three quarters; tarsus nine tenths of an inch; beak, from gape, three fifths of an inch.

This pretty little Warbler, the smallest of the European Reed Warblers, is at once distinguished from all others by its curved beak and fan-shaped black and white tipped tail. It was first described by Temminck, in the first edition of the "Manual," from skins brought to him from Portugal, by M. M. Link and Hoffmannsegg, and was subsequently taken by Natterer in some plenty at Algesiras, in the neighbourhood of Gibraltar. It belongs to the genus *Cysticola* of Lesson and Bonaparte, and is closely allied to the so-called Beutelsingers which inhabit especially Asia, Africa, and New Holland, bearing as Count Mühle observes, the same relation to the other Warblers as the Beutel Titmouse does to the other Titmice.

It is found in Portugal, the marshes near Rome, Tuscany, Sardinia, and Sicily, where it is very common. In France it is principally found on the banks of the Var, and the marshy country of Camargue. It is found along the whole shore of the Mediterranean, is plentiful in Greece, and extends even to the south of Siberia.

Count Mühle remarks, "that it appears always lively and cheerful, winter or summer. It loves to live among sedges and rushes in ponds and swamps, and may be often seen rocking itself with evident pleasure on the top of the papyrus plants, *(Cyperus papyrus.)* In summer it dwells by choice in swampy grounds, and when these become cold and bare in winter it resorts to the high grass of the meadows and corn-fields. It does not appear to frequent the cistus tree, and therefore the name given to it by the Germans, *Cistensänger*, is misapplied."

In its habits we perceive a strong similarity to the rest of its family. "If it hide among the grass, in a few minutes it will be perceived coming up to higher branches of the shrubs, fluttering and hopping after smaller insects on the leaves of the sedges and reeds, and, rising suddenly in the air, stop a short time and then come down to the same spot, from whence it will again conceal itself among the bushes."

Its flight is not rectilineal, but takes the form of a curve, which corresponds to the repeated flapping of its wings, during which it frequently utters its sharp shrill call-note. This resembles much that of *Anthus pratensis*, and is its entire song. It is heard loudest when it thinks its nest is in danger, upon which occasions it wheels round the intruder's head in circles, uttering loud cries.

"It breeds three times in the year: the first time in April, when the nest is built negligently, for want of materials, and it generally chooses a sedge or rush clump for its home. It is always one foot distant from the ground. The perfect nest is a very beautiful and wonderful construction; the stalks of the plants which form the outside are entwined together with a perfect needle-work, not merely strung one to another. In the sides of every leaf the little bird reaches with its beak, it makes a small opening, and passes through plant fibre threads, formed from the *Asclepiadæ*, *Epilobiæ*, and the pappous of *Syngenesia*. Those threads are not very long, and reach only twice or thrice from one leaf to another, and it is astonishing how these little birds so elegantly and solidly accomplish the troublesome work. The inside of the nest is merely spread over with the down of plants. The eggs, in number from four to six, are more or less a lively greenish grey; and, according to Savi, some are flesh-coloured."

Brehm, in Bädeker's European eggs, calls this bird

the "European Tailor Bird," from the manner in which it sews up the materials of its nest: he declares also that a knot was found by Herr V. Kœnig at the beginning of the thread! The narrow entrance is either on the side or in the top, so that the nest has the form of a bag. In the inside he says it is lined with long grass leaves, and the eggs are five or six, in size between our Common Wren and Goldcrest; they have a very soft shining shell, and are either blue green, like the Redstart, or paler, like those of the Pied Flycatcher; or they are sometimes of a shining white only.

M. Moquin-Tandon has kindly sent me drawings of three varieties of the egg, from which my figures are taken: they were accompanied by the following remarks: —"*Sylvia cisticola*, Savi, from the south of France. This bird's very remarkable nest is now well known, and has been drawn many times,—those of Schinz and Roux are sufficiently exact. I have lately given a detailed description of this charming bird. It is well known that the bird sews its nest to a tuft of carices or grasses. Its form is that of an oblong purse, which opens obliquely towards the top. It lays four or five, and sometimes six, whitish eggs, which sometimes approach to very clear blue, (A,) and sometimes a rosy tint, (B.) M. Le Brun sent me one variety much darker."

The male and female specimens sent me by M. E. Verreaux, have the upper parts strongly marked with black in the centre of the feathers, with a shade of russet and grey on the borders; head and neck varied with the same colours, and the rump red; throat and middle of the belly white; the crop, sides of neck, flanks, and under tail coverts russet yellow. Wings like the back; the primaries with less black in the centre of the feathers; tail brownish black, the ends of the quills tipped with ash-colour; beak, (which is curved,) feet, and iris brown.

According to Savi the interior of the beak of the female during breeding time is yellow, that of the male violet black.

The young before the first moult resemble the old birds, only the spots on the upper part of the plumage are less, and not so dark.

My figure is from a specimen sent me by M. E. Verreaux. Figured also by Temminck and Laug, pl. col. 6, f. 3; Roux, Ornith. Prov., pl. 232; Gould, B. of E.; Mühle, Monograph, in two plumages, one the normal type described above, the other having no spots or stripes on the vertex. The specimen was brought from Greece by Count Mühle, and is exactly the same as those described by Temminck from Japan.

INSECTIVORÆ.
Family SYLVIADÆ.
Genus SYLVIA. *(Latham.)*

CETTI'S WARBLER.

Sylvia cetti.

Sylvia cetti,	MARMORA; Mem. della Acad. di Torino, vol. xxv., p. 254.
" "	TEMMINCK; Man., 1820.
" "	SCHINZ; Europ. Faun., 1840.
" "	Z. GERBE; Mag. de Zool., 1840.
" "	MÜHLE; Monog., 1856.
" *sericea,*	NATTERER.
Cettia altisonans et sericea,	CH. BONAPARTE, 1838.
" " "	Z. GERBE; Dict., 1848.
" *cetti,*	DEGLAND, 1849.
Salicaria cetti,	KEYSERLING AND BLASIUS; Der Wirbelthiere, 1840.
" "	SCHLEGEL; Revue, 1844.
Bec-fin Bouscarle,	
Bec-fin Cetti,	OF THE FRENCH.
Cetti's Rohrsänger,	OF THE GERMANS.
Usignuolo di Fiume,	OF CETTI; in Ucc. di Sardegna.

Specific Characters.—Upper parts of the body unicolorous rich dark chesnut brown; superciliary ridge whitish, long, and narrow; the tail rounded, consisting of only ten quills; tarsi clear brown. First primary of medium length, second equal to ninth, the third shorter than the fourth, fourth and fifth longest.

Dimensions of specimens sent me by M. E. Verreaux.—Male:— Length from tip of beak to end of tail five inches; carpus to tip two inches and a half; tail two inches; beak three fifths of an inch; tarsi nine tenths of an inch. Female:—From tip of beak to end of tail four inches and a half; carpus to tip two inches; tail one inch and three quarters; beak three fifths of an inch; tarsi eight tenths of an inch.

CETTI's Warbler, distinguished from all other *Sylviadæ* by having only ten quills instead of twelve in the tail, is found in the whole of Southern Europe, from Spain to the Caucasus. It has been said to have been killed, but erroneously, in England, and Count Mühle, in expressing his surprise at this, attempts to account for it by attributing it to a deficiency in the development of the wings,—a reason which I think can hardly be maintained, when we hear of much weaker birds flying an infinitely greater distance. It occurs in Sicily, Corsica, Sardinia, Spain, and France, being especially common in winter in the southern provinces of the latter country. M. Gerbe reports its appearance in the department of the Var, and M. Crespon indicates many localities in which it is found in Provence; and Count Mühle found it in Greece. It occurs in Egypt, and, according to Captain Loche, in the three provinces of Algeria. Mr. Salvin says, in the "Ibis," for July, 1859:—"On one or two occasions, among the tamarisk trees on the banks of the Chemore, I caught a momentary glimpse of a bird of this species—not more than was sufficient to recognise it. It appears to be shy and not common in the Eastern Atlas of Africa."

It is only found in the thickest and most impenetrable coverts of grassy plants, and thick hedges and ditches. It is very shy, more so than any other Reed Warbler: should it chance to climb up on a branch or reed-stem, it is down again the moment it is observed. Its song is rather agreeable, but is heard only from its concealment. The poor bird seems to have a more than usual instinctive knowledge that reasoning man is its enemy. Its call of two syllables resounds continually. When it is pursued, and it thinks the enemy has been led away far enough, it will turn quickly back again to its first place.

It builds its nest in a bush not far from water, and near the ground. It is constructed of dry grass and half-decayed plant stems: it lines it with horse-hair and sometimes with feathers. It lays four or five eggs, which are brown red, without spots, and as large as those of the Whitethroat.

My figure of the egg is from a drawing sent me by M. Moquin-Tandon, who remarks about its nidification: —"*S. cetti*, from the neighbourhood of Montpellier, where the bird is rather rare. This egg was given me by M. Devilliers. *S. cetti* makes its nest in bushes or large aquatic plants, at a short distance from the ground: the nest, skilfully made, is composed of stalks of grasses, and also of feathers. It contains four or five eggs of an uniform red brick-colour, without spots. It is sometimes darker than the drawing. I saw lately at a Paris merchant's ten eggs of this bird: six of them like the drawing, three darker, and one lighter. Great diameter nineteen to twenty-one millimetres, small diameter fourteen or fifteen."

The male bird sent me by M. E. Verreaux has all the upper parts of the body a rich chesnut brown, darkest on the wing primaries and the tail. The throat is white, shading off to ash grey on the belly and to olive brown on the flanks and under tail coverts, the latter being tipped with white. The wings are short, only just covering the rump. Beak and feet light brown; iris nut brown.

The female has the colours slightly paler than the male, but they are difficult to distinguish, except by size. Young before the first moult are, according to Degland, of a darker brown than the adult.

My figure is from specimens sent me by M. E. Verreaux. Figured also by Buffon, pl. enl. 655, under the name of *Bouscarle de Provence;* M. Roux, Ornith. Prov., pl. 212; Z. Gerbe, Mag. de Zool., 1840, pl. 21; Gould, B. of E.

INSECTIVORÆ.
Family *SYLVIADÆ.*
Genus Sylvia. *(Latham.)*

RIVER WARBLER.

Sylvia fluviatilis.

Sylvia fluviatilis,	Meyer and Wolff; Tasch. des Deuts, (1810,) tome i., p. 229.
" "	Naumann, tome iii., 649.
" "	Temminck; Man., (1820.)
" "	Vieillot; Dict., 1817.
" "	Schinz; Europ. Faun., (1840.)
" "	Mühle, 1856.
Locustella fluviatilis,	Gould; B. of E., 1836.
" "	Bonaparte, 1838.
" "	Z. Gerbe; Dict., 1843.
" "	Degland, 1849.
Salicaria fluviatilis,	Keyserling and Blasius, 1840.
" "	Schlegel; Revue, 1844.
Riverain Fluviatile,	Of the French.
Fluss Rohrsänger,	Of the Germans.

Specific Characters.—Above unicolorous olive brown, darker on the wings; under tail coverts light rust grey, with white at the end; first primary the longest; wings reach to half the length of the tail.

Length of a male sent me by M. E. Verreaux, six inches; carpus to tip three inches; tail two inches and one fifth; tarsus nine tenths of an inch; beak seven tenths of an inch.

The River Warbler, which is one of the finest species in the family, is found principally in Europe on the shores of the Danube. It has also been found, but isolated and rarely, in Saxony, Siberia, Lithuania, and France. It also occurs in Hungary and in Egypt. Its home is in moist and swampy places, where reeds, high grasses, and water-plants afford it concealment. It is migratory, appearing in its breeding places on the banks of the Danube in May, and disappears the end of August. Of its habits, Count Mühle says,—

"By day it keeps in the deepest concealment, and flies away when disturbed with the greatest stillness and velocity, or it hastens from one bough to another, close over the ground. In early morning, however, in the still gloomy twilight, it will remain quite unconcerned, singing on an open branch or twig, and even by day it may sometimes be heard during thunder-storms. While singing, it likes to sit upon a slanting branch, swells out its throat, lets its wings droop somewhat, and with a measured movement sings its remarkable song in quick railroad time, repeated quite twenty times in a breath. This song resembles the chirping of grasshoppers. Upon the very obscure history of its propagation some light has been thrown by Thienemann and Heckel. Its breeding-places are the closely-wooded parts of the banks of the Danube. The nest is always in thick bushes, which have growing amongst them high grasses and reeds. It is formed of withered leaves, mostly of grass, and dry straws, thickly woven over with the young shoots of grasses, so as to conceal it completely

from observation. The inside of the nest is cup-shaped, and neatly and solidly lined with small soft grass stems, without any mixture of other materials. The four eggs which Heckel found in one such nest, (Naumannia, p. 17, 1853,) have a conspicuous greyish white ground, with reddish brown spots, some light, others dark, and slight stains scattered over."

Brehm, in Bädeker's work upon European eggs, gives a still later account of its nidification, which I will quote entire, as everything about this bird is interesting to the naturalist:—

"It dwells, but not numerously, in the high-lying meadows of the Elbe, by Magdebourg and Breslau, and it is plentiful on the shores of the Don, the Bug, and many rivers in Gallicia. It lives in the woods and thick bushes on the banks of the rivers. It breeds, according to Count Wodyccki, in Gallicia, and to others in Moldavia, not far from Prague, and on the Elbe. Herr Zelebor shot a female with an egg just ready to lay, May 22nd., 1852, and found the nest on the Don not far from Vienna. It builds in bushes which are thickly grown through with reed-grass and *Parietaria officinalis*. The nest is formed of dry reeds and grass leaves, tender twigs, strong grass stems, strips of reed, etc., interwoven with dry meadow grass. It is lined with soft grass. The eggs are four or five, which are greyish, inclining to reddish ground-colour, upon which are indistinct pale violet-grey and darker or lighter spots and streaks of reddish brown, thicker at the base. They are unequally shaped, gently declining from the base to the top."

The River Warbler feeds on insects and flies.

The male in breeding plumage has all the upper parts of the body olive green, shaded with brown; the throat white; under wing and tail coverts, and all the under parts of the belly whitish, shaded with light olive green; the anterior part of the neck mottled with olive green and ash-colour. Primaries rich brown; first the longest, and the others gradually a little less down to the ninth, giving the wing a long pointed character. This graduated position of the primaries is shewn when the wing is in repose. The beak is dark brown above, light below. Tarsi light yellow; iris, dark brown.

The female has the upper parts like the male; the throat, neck, and crop dirty white, feebly marked with long spots of ashy brown. In autumn the feathers are bordered with ash.

My figures of this bird and its egg are from specimens sent me by M. E. Verreaux.

Figured also by Naumann, Taf. 13; Gould, B. of E.

INSECTIVORÆ.
Family SYLVIADÆ.
Genus SYLVIA. *(Latham.)*

PALLAS'S LOCUSTELLE.

Sylvia certhiola.

Motacilla certhiola,	PALLAS; Zoog. i., p. 509, No. 141.
Sylvia certhiola,	TEMMINCK; Man. i, p. 187.
Calamoherpe certhiola,	BLASIUS; Naumannia, 1858.
Bec-fin Trapu,	OF THE FRENCH.

Specific Characters.—Beak strong; superior mandible black; the plumage of the upper parts varied with numerous spots; all the quills of the tail are terminated below by a large ashy blotch; posterior claw much curved, and longer than the toe. Length five inches.

THIS bird, closely allied to the Grasshopper Warbler, so well known to the naturalists of Great Britain, was introduced into the European Fauna by Pallas, in his "Fauna Rossica," as *Turdus certhiola,* and was described by him afterwards under the name of *Motacilla certhiola.* It was subsequently introduced into the Manual by Temminck as *Sylvia certhiola,* who has given a very clear description of the bird, and a careful comparative diagnosis between it and *locustella.*

It was however erased from the European list by Schlegel, upon very much the same kind of evidence which induced that distinguished ornithologist to deal in like manner with *Falco leucocephalus*, and it is left out by Degland upon this authority.

It has however recently come to light in that singular island Heligoland, where so many new birds, especially American forms, have been added to the European list. It was found by Herr Gätke, and has been introduced into the list of Birds new to Europe, described by Professor Blasius, in Naumannia, for 1858, p. 311, in the following words:—"*Calamoherpe certhiola* is a perfect gem in its new plumage; and here for the first time found in Europe. There are two specimens mentioned by Middendorf, as having been killed on the shores of the Sea of Ochotzk. Before these specimens were found, the original from Pallas, in the Berlin Museum, was the only one known."

Of this rare bird then there are it appears only four specimens in existence, a fact which is perhaps sufficiently accounted for by its similarity to the Grasshopper Warbler, whose well-known retiring habits fortunately protect it from wholesale destruction, and make it even a rare bird. *S. certhiola* is no doubt closely allied to *locustella*, both in organization and habit. It is only distinguished by its stronger beak, by the posterior claw being in *certhiola* longer than the toe, and more curved, while in *locustella* the claw is shorter than the toe. Further, the tail of the latter is unicolorous; while in the former, as seen by the specific characters at the head of this notice, each tail quill is terminated below by a large ashy blotch. In all other respects the disposition of colours in the plumage of the two birds is the same.

S. certhiola has, according to Temminck, whose description I follow, all the upper parts of a uniform olive, shaded with brown and varied by ovoid spots of blackish brown; these spots occupy the middle of each feather; the throat, front of neck, and middle of the belly pure white; under the throat is a zone of very small ovoid spots of dark brown; flanks, abdomen, and inferior tail coverts of a bright russet, the last-named tipped by pure white; tail long, wide, and much sloped. The quills are blackish below, and all terminated by a large mark of whitish ash-colour; but *above* there is only a small point of the quills in which this ashy mark is perceptible.

The female has the colours less marked and pure.

Figured by Gould, pl. 105, from which my drawing is by permission taken.

With this bird I close my list of the European species of the genus *Sylvia* of Latham. The progress of ornithological discovery in modern days, however, renders it probable that the number will be considerably increased, as every single well-authenticated case of the capture of a bird within the European limits is held sufficient to constitute it a European species by modern writers. I hope to see this system some day altered by the multiplication of such excellent memoirs as those of Tristram and Salvin in the Ibis, by which our knowledge of geographical ornithology will be much increased, and our boundary species placed in their respective habitats. It must at the same time be borne in mind as our geographical divisions are entirely arbitrary, so is it impossible to draw a distinct line between the species of one quarter of the globe and another, and yet the faunæ are sufficiently distinct to afford a remarkable illustration of the adaptation of species to climate and country, and hence a confirmation of the doctrine of special creation, and permanence and immutability of species—a great truth which has lately been assailed in a work which contains more sophistry and unsound deduction than any book that has ever been printed.

Of the species of *Sylviadæ* omitted in the present work, I will make one or two remarks:—

1. *Sylvia lanceolata*, of Temminck.—This was said to have been discovered in the neighbourhood of Mayence. This however has been denied by Bruch, who gives the South of Russia as the locality of the only two specimens said to have been taken in Europe. Brehm considers Temminck's specimens were those of *locustella;* Mühle that there is no sure foundation for the identity of the species. Malherbe considers it has been prematurely admitted into the European list. Degland and Schlegel recognise a specimen killed at Genes, by the Marquis Durazzo, as evidence of its claim. On the whole I have thought it better to omit this bird.

2. *S. familiaris*, Menetries.—Considered by Naumann as identical with *S. galactodes*, which opinion is denied by Schlegel. Count Mühle, who had many opportunities for examining this so-called species in Greece, says that they are undoubtedly different in the colouring of the upper part of the body. Temminck and Keyserling and Blasius think them identical. Degland, in a note to *S. galactodes*, merely draws attention to Schlegel's statement.

The Rev. H. B. Tristram, who has had many opportunities of observing this bird, both in Western Africa and in the Levant, considers the species identical, and states that he has found greater difference in the colouring of the back part of the body, between individual Algerian specimens, than between the ordinary African and Greek types.

3. *S. borealis*, (*Phyllopneuste borealis*, Blasius,) is a new species, intermediate between the *P. javanica* of Horsfield, and *S. ieterina* of Vieillot. The following is Professor Blasius' account of this bird, from Naumannia, 1858. I had intended to have deferred any account of the species until the end of the work, but as it may by that time lose its present claim to specific distinction, I will give a short abstract of Professor Blasius' description. The bird was killed in Heligoland:—

This bird forms with *P. javanica* a natural group under the Foliage Warblers. Upper part of the body and the edges of the wing and tail feathers yellowish grass green; vertex gradually darker coloured, grey green. The tail feathers slender, and have their greatest breadth a fourth from base; the inner web here turns in a disproportionate manner towards the shaft. The whitish tips of the first tail feathers are broadest near the point of the shaft on the inner web. In the above characters it differs entirely from *P. javaniea*. The under part of both species is white along the middle, with a faint brimstone yellow flush. Feathers of the sides of the head and the front part of the neck are on the tips and edges clouded with grey, while in *P. javanica* these parts are clear yellowish white. The flanks shade off into the colours of the back, with greenish grey. The small first primary is only a little longer than the upper wing coverts, third and fourth form point of wing, fifth larger than sixth, smaller than second. Tail tolerably straight, two middle feathers somewhat lengthened; beak dark horn blue colour, with yellow flesh-coloured edges. Feet bright-coloured in *javaniea*, brown grey; in *borealis* bluish green.

	S. javanica.		S. borealis.		S. icterina.	
DIMENSIONS.	Inch.	Lines.	Inch.	Lines.	Inch.	Lines.
Length of wing	2	3.2.	2	5.4.	2	6.0.
Length of tail	1	7.8.	1	8.8.	1	11.5.
Head with beak	1	1.3.	1	2.2.	1	0.5.
Beak from nasal orifice	0	3.4.	0	3.6.	0	2.6.
Tarsi	0	9.0.	0	9.0.	0	8.8.
1st. wing feathers longer than the upper coverts	0	3.0.	0	1.0.	0	4.0.

4. *S. pinetorum*, Brehm; 5. *S. arbustorum*, Brehm; and 6. *S. hydrophilos*, are considered by Brehm to differ from the Reed and Sedge Warblers, to which they are respectively closely allied.

7. *S. moussieri* has been stated to have occurred in Europe by Herr L. Alph Galliard, but this is discredited by Mühle; it has however been introduced by Dubois. Mr. Salvin states (Ibis, vol. i., p. 307,) that it is exclusively a North African species.

Mr. Tristram remarks that this species is a very local one in Africa, never, in Algeria, crossing the Atlas, and in Tunis only approaching the coast at the south-west corner of the Gulf of Cabes. It is chiefly an inhabitant of the Oases, where it is a constant resident.

8. *S. ochrogenion*, Lindermayer, (Isis, 1843,) is the female of *S. melanocephala;* the yellow colouring observed on the throat is considered by Count Mühle to be produced by the fruit of the *Cactus opuntia*.

9. *S. sylvestris*, Naumann, (App. xiii., p. 429, table 369.)—Under this head Count Mühle enlarges upon the difficulty of separating many forms of the Willow Wrens, and hints at the possibility of this proving a fifth species.

10. *S. horticola*, Brehm, (Naumann, xiii., p. 444.)— Once more, says Count Mühle, we have to deal here with a bird out of the great Brehm sub-species manufactory.

11. *S. fruticola,* Naumann, (xiii., p. 453.)—Supposed to be the young of *S. nigrifons,* itself a not established species. It is with the last evidently a variety of *S. arundinacea* or *S. palustris.*

12. *S. rubricapilla,* Landbeck, *S. naumanni,* Müller.—Has been determined as a species by Müller, who refers it to the above figure in Naumann's work, xiii., p. 411. It is distinguished from *atricapilla* by the male having a red head like the female. The species has not been satisfactorily determined.

13. *S. obscurocapilla—Calamoherpe obscurocapilla,* Dubois.—Is a species figured, with its nest and eggs, and described as new by Dubois, in his beautiful work, which he has kindly sent to me, the "Birds of Belgium." It is quite possible that this species will prove distinct, though in the absence of specimens for examination I do not feel competent to give a stronger opinion. It appears to have been first discovered by M. le Vicomte de Spoelberg, in 1854. It is, says M. Dubois, closely allied to the Marsh Warbler, *(S. palustris,)* with which in fact it has long been confounded. From this however it differs in the male having a dark head, which is never observed in *palustris.* It appears to be a great mimic. The nest is similar to that of the Garden Warbler in form, and of the Grasshopper Warbler in construction.

14. *S. arigonis. Hypolais arigonis,* A. Brehm, *fils. Algemeinen deutschen naturhistorischen Zeitung,* No. 161, p. 467, tome 3.—This is said to be a good species by M. Dubois. If it maintains this character I shall endeavour to give a notice of it in the Supplement.

The above include all, I believe, of the real or supposed and doubtful species, which have been thought by various authors to have had more or less claim to admission into the European list of birds.

INSECTIVORÆ.
Family SYLVIADÆ.
Genus REGULUS. *(Ray.)*

Generic Characters.—Beak very slender, short, straight, dilated at the base, and compressed towards the point; edge of the mandible slightly turned in; nostrils oval, covered by two small, stiff, arched feathers, with slightly disunited webs. Wings of moderate length, the first quill very short, the second and eighth equal, the fourth and fifth the longest in the wing. Legs slender; the middle toe united at its base with the external, the posterior the strongest of all; tail sloping, and composed of ten feathers.

RUBY-CROWNED KINGLET.

Regulus calendula.

Motacilla calendula,	LINNÆUS.
Regulus calendula,	LICHTENSTEIN. AUDUBON.
" "	WILSON. BONAPARTE.
Sylvia calendula,	NUTTALL; Man., vol. i., p. 415.
Le Roitelet rubris,	BUFFON, 373.

Specific Characters.—General plumage like that of the other Goldcrests. The silky feathers of the crown of the head vermilion; under parts greyish white. Length four inches and a quarter; wing six inches; beak one third of an inch; tarsus three fourths of an inch.

The introduction of this beautiful and very distinct species of one of the smallest birds of the new world into the European and British fauna, may perhaps excite a smile of incredulity in some of my readers. But the facts connected with its capture in the Scotch Highlands are conclusive, and cannot admit of doubt.

During my residence in Edinburgh, in 1859, I received a letter from Mr. Robert Gray, of Glasgow, informing me that the "Ruby-crested Wren" had been shot on the banks of Loch Lomond, by Dr. Dewar. The original specimen had been presented to Mr. Gould, but Mr. Gray kindly sent me an accurate and beautiful drawing of the bird, made by Mr. Sinclair, which I have much pleasure in introducing into my work. A notice of its capture I had the pleasure of giving at one of the meetings of the Physical Society.

Having quite satisfied myself by repeated correspondence, that there could have been no mistake about the matter, I shall content myself by introducing here part of Dr. Dewar's letter. There may, I think, be some truth in Dr. Dewar's suggestion, that these small birds get a lift *en route* in the numerous vessels which are constantly passing between the two countries. But after all there is nothing very extraordinary in such a migration. The little creature is in America a *migratory bird*, and flies, according to Audubon, from Louisiana and other southern states to Newfoundland and Labrador, where it breeds. It leaves the south in March, and has young in the far north in June. Our friend had evidently mistaken its way back again to the south, and come to the Scotch Highlands instead.

The following is an abstract from Dr. Dewar's letter, dated Glasgow, December 6th., 1859:—"The specimen of *Regulus calendula*, regarding which you write, I shot in Kenmore Wood, Loch Lomondside, in the summer of 1852. Living in its vicinity, I went to the wood for the purpose of shooting some specimens of Goldcrests, which are always there in abundance. After procuring upwards of a dozen, I found, on looking them over, what I took to be the Firecrest: this I safely deposited among my other skins, where it lay till last year, when, on examining it carefully, with the view of exhibiting it at the Natural History Society here, to my surprise my specimen turned out to be, not *Regulus ignicapillus*, as I had supposed, but *Regulus calendula* of North America. I forwarded it to Mr. Gould for examination, to whom I afterwards presented the specimen. Although I look upon the occurrence of *Regulus calendula* in this country as a subject of extreme interest, still it has no claim to a place among our birds, farther than as one of the many stragglers which from time to time find their way to our shores. How this little creature, the most diminutive of all the American species which have visited Britain, found its way across the Atlantic is almost inconceivable. My belief is that most of the American species which are met with in this country, are aided in their passage by vessels crossing the Atlantic, and I think it utterly impossible for such a tiny bird as this to find its way across without some such assistance. Two or three instances have occurred to my own observation, in which birds were conveyed in this way."

Audubon's account of this bird is so interesting that I will take the liberty of making from it a very long extract. His writing is always welcome to the real lover of nature.—"The history of this diminutive bird is yet in a great measure unknown, and although I have met with it in places where it undoubtedly breeds,

I have not succeeded in finding its nest. On the 27th. of June, 1833, while some of my party and myself were rambling over the deserts of Labrador, the notes of a Warbler came on my ear, and I listened with delight to the harmonious sounds which filled the air around, and which I judged to belong to a species not yet known to me. The next instant I observed a small bird perched on the top of a fir tree, and on approaching it, recognised it as the vocalist that had so suddenly charmed my ear and raised my expectation. We all followed its quick movements as it flew from the tree backwards and forwards, without quitting the spot, to which it seemed attached. At last my son John raised his gun, and on firing brought down the bird, which fell among the brushwood, where we in vain searched for it.

The next day we chanced to pass along the same patch of dwarf wood in search of the nests of certain species of ducks, of which I intend to speak on another occasion. We were separated from the woods by a deep narrow creek; but the recollection of the loss of the bird, which I was sure had been killed, prompted me to desire my young friends to dash across and again search for it. In an instant six of us were on the opposite shore, and dispersed among the woods. My son was so fortunate as to find the little *Regulus* among the moss near the tree from which it had fallen, and brought it to me greatly disappointed. Not so was I, for I had never heard the full song of the Ruby-crowned Wren, and as I looked at it in my hand I could not refrain from exclaiming, 'And so this is the tiny body of the songster from which came the loud notes heard yesterday.' When I tell you that its song is fully as sonorous as that of the Canary bird, and much richer, I do not come up to the truth, for it is not only as powerful and clear, but much more varied and pleasing to the ear. We looked for its mate and nest, but all around was as silent as death, or only filled with the hum of myriads of insects. I made a drawing of it in its full spring plumage. A month later the young of this species were seen feeding among the bushes.

The Ruby-crowned Wren is found in Louisiana and other southern states from November to March. Near Charlestown they are sometimes very abundant. The old birds are easily distinguished from the young without shooting them, on account of the curious difference in their habits, for while the latter keep together among the lowest bushes, the former are generally seen on the top branches of high trees. I have not observed a similar difference in *Regulus tricolor*. The rich vermilion spot on the head in the parent species was wanting in the young, that part being of the same plain colour as the back.

I have found this bird in Kentucky also during winter, but generally in southern exposures, and usually in company with the Brown Creeper and Titmouse.

The little bird of which I speak appears to feed entirely on insects and their larvæ, and I have often thought it wonderful that there should seem to be no lack of food for it even during weather sometimes too cold for the birds themselves. It seems to migrate during the day only, and merely by passing from bush to bush, or hopping among the twigs, until a large piece of water happens to come in its way, when it rises obliquely to the height of above twenty yards, and then proceeds horizontally in short undulations. It emits a feeble chirp at almost every motion. So swiftly, however, does it perform its migrations from Louisiana to Newfoundland and Labrador, that although it sometimes remains in the first of these countries until late in March, it has young in the latter by the end of June, and the brood is able to accompany the old birds back to the south in the beginning of August.

The adult male in summer plumage has the bill short, straight, subulate, very slender, and compressed, with inflated edges; upper mandible nearly straight in its dorsal outline, the edges scarcely notched close upon the slightly declinate acute tip; lower mandible straight and acute. Nostrils basal, elliptical, half closed above by a membrane covered over by feathers. The whole form is slender, although the bird looks somewhat bulky, on account of the loose texture of the feathers. Legs rather long; tarsus slender, longer than the middle toe, much compressed, and covered anteriorly with a few indistinct scutella; toes scutellate above, the lateral ones nearly equal and free, hind toe stouter; claws weak, compressed, arched, and acute.

Plumage very loose and tufty. Short bristles at the base of the bill. Feathers of the head elongated and silky. Wings of ordinary length, third and fourth primaries longest. Tail of twelve feathers, emarginate, and of ordinary length; bill black, yellow at the base of the lower, and on the edges of the upper mandible. Iris light brown; feet yellowish brown, the under parts yellow. The general colour of the upper parts is dull olivaceous, lighter behind. The eye is encircled with greyish white, of which colour also are the tips of the wing coverts; quills and tail dusky, edged with greenish yellow; the silky feathers of the crown of the head vermilion. The under parts greyish white.

Length four inches and a quarter; extent of wings six inches; bill one third of an inch; tarsus three fourths of an inch.

The adult female in summer plumage resembles the male, but the tints are in general duller, especially the greenish yellow of the wings.

Figured by Audubon and Wilson.

Since the above was in type I have been informed by the Rev. H. B. Tristram, that he has a specimen of this bird which he had in the flesh, and which was killed by a Durham pitman in 1852, in Branspeth Woods.

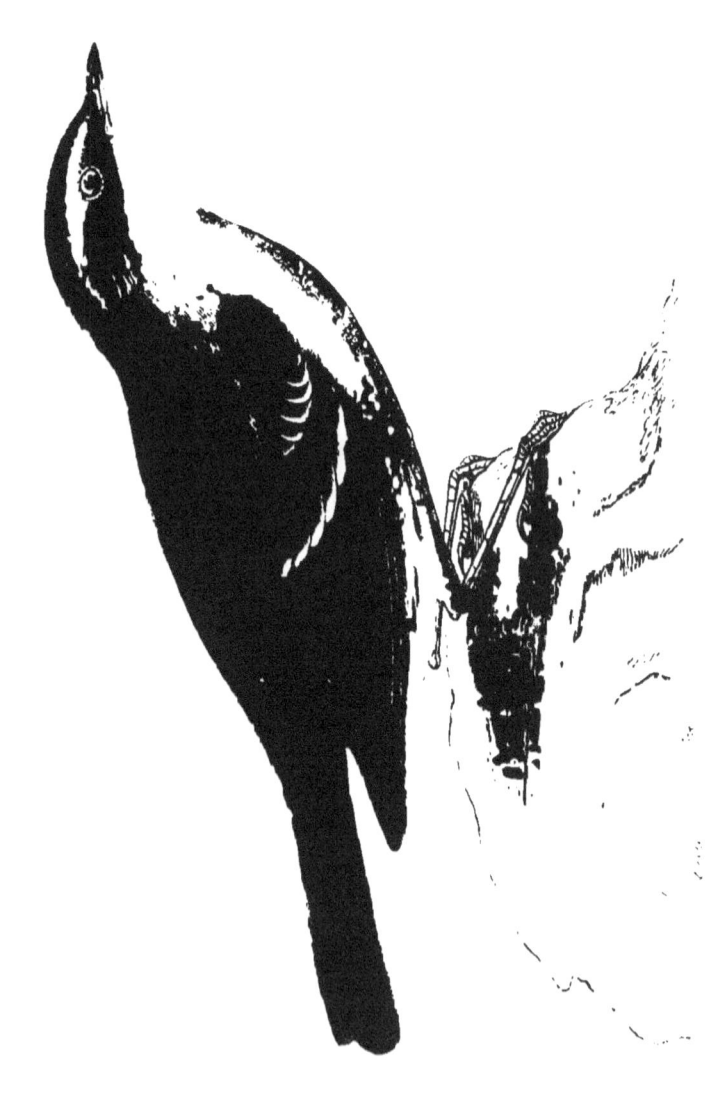

INSECTIVORÆ.
Family SYLVIADÆ.
Genus ACCENTOR. *(Bechstein.)*

Generic Characters.—Beak of medium length, robust, straight, conical, and pointed. Edges of each mandible compressed, upper one notched near the point. Nostrils basal, naked, pierced in a membrane of considerable size. Feet robust, three toes in front and one behind, the exterior joined at its base to the middle toe; the claw of the hind toe by much the longest and the most curved. First wing primary nearly obsolete, the second nearly as long as the third, which is the longest in the wing.

MOUNTAIN ACCENTOR.

Accentor montanellus.

Motacilla montanellus,	PALLAS; Voyage, table 8, French Edition, Appendix, p. 71, 1776.
Accentor motanellus,	TEMMINCK; Man., 1820.
" "	BONAPARTE. SCHINZ.
" "	KEYSERLING AND BLASIUS.
" "	SCHLEGEL. DEGLAND.
Accenteur Montagnard,	OF THE FRENCH.
Berg Fluhvogel,	OF THE GERMANS.

Specific Characters.—A double row of transverse yellow spots on the wing. Head above deep black; shafts of the tail quill feathers russet. Length about six inches.—DEGLAND.

This rare and interesting bird is an inhabitant of the south-east of Europe, being found principally in Siberia, Russia, and the Crimea. It occurs, but accidentally, in the south of Hungary, in the Neapolitan States, and Dalmatia. Its range in Asia is along the same latitude as in Europe. Mr. Tristram informs me that the only specimen he ever saw was in the Museum at Palermo, and he at the same time directed my attention to Middendorf's account of his capture of this bird. I copy the following from his "Sibirische Reise," vol. ii., p. 172:—

"One specimen only of this bird was shot in the Stanowój chain of mountains. It perfectly agrees with Pallas's description and Gould's painting, and consequently entirely removes the doubt which Brandt expressed relating to the identity of the Siberian and European Birds, as well as to that which received the name of *Accentor temminckii*. The specimen mentioned by Brandt is in the winter plumage; therefore being in a transition state, it is fainter in colour, and greyish. The stripe over the eyes is white; the throat dirty white; breast and belly bright rust yellow. On the back of the bird the colours are an admixture of rust brown and ash grey, which latter predominates, especially about the head and nape. The spots have the same colour as in the summer plumage, but in the latter they extend more towards the wings. The Accentor, *Atro-gularis* of Brandt, differs from *A. montanellus* essentially in the black throat."

The following is Temminck's description of this bird: —"The adult male has a hood of deep black, covering the head and occiput; a large equally black band passes below the eyes, and covers the orifices of the ears; a large yellow eyebrow takes its origin at the

EGG OF ACCENTOR MONTANELLUS.

I HAVE much pleasure in giving here a figure of the egg of *Accentor montanellus*, kindly sent me by Professor Moquin-Tandon, with the following remarks: —"My two eggs of this bird are exactly alike in shape and colour; they are twenty-three millimetres in long diameter, and sixteen in the short. The colour pale and uniform azure blue. They were taken in the south of Hungary, and sent to me by my friend M. Raoul de Baracé D'Angers."

base of the beak, and is continued to the nape; the upper parts of the body and the scapularies are of an ashy russet, marked with large longitudinal spots of a brick red. Wings of an ashy brown, bordered with grey russet; two rows of small yellow points form on the wing a double band; tail of a unicolorous brown, but the feather shafts of a russet brown. All the inferior parts are of an isabel yellow, varied on the crop with brown spots, and on the flanks with longitudinal spots of a grey russet; base of the beak yellow, point brown; feet yellowish. Length five inches three or four lines. The female is of a blackish brown on the head, on the occiput, and on the auditory orifices. It does not otherwise differ from the male."

The Mountain Accentor is stated to feed in the spring upon the same kind of food as its congener, our well-known old friend the Hedge Sparrow. In winter Temminck thinks it may be guilty of feeding upon seeds. Of its propagation I find authors are silent.

My figure is taken by permission from Mr. Gould's B. of E.

INSECTIVORÆ.
Family SAXICOLINÆ.
Genus Saxicola. *(Bechstein.)*

Generic Characters.—Beak slender, straight, and rather broader than deep at the base, where it is surrounded by a few hairs; upper mandible slightly obtuse, sloped, and curved only at the point; nostrils oval, half-closed by a membrane; tarsus long, slender, and compressed; outer toe connected by a membrane to the middle. Wings of medium length, reaching to the middle of the tail when closed; first quill about half as long as the second, the second shorter than the third or fourth, which are the longest; the greater wing coverts much shorter than the quills. Tail medium length, slightly rounded or square, consists of twelve quills.

BLACK WHEATEAR.

Saxicola leucura.

Turdus leucurus,	Gmelin; Syst., 820, 1788.
" "	Latham; Ind., 1790.
Œnanthe leucura,	Vieillot; Dict. et Faun. Fr., 1818.
Saxicola cacchinans,	Temminck; Manual, 1820.
" "	Schinz. Schlegel.
" *leucura*,	Keyserling and Blasius; Die Wirbelt, 1840. Mühle.
Vitiflora leucura,	Bonaparte, 1838.
Dromolica leucura,	Cabanis. Bonaparte.
" "	Tristram; (Ibis.) Loche.
Traquet Rieur,	Of the French.

Schwarzer Steinschmätzer,   Of Meyer.
Culbianco Abbrunato,   Of Savi and Marmora.

Specific Characters.—Plumage black, or blackish, with the upper and lower tail coverts white; tail white, with half of the two middle quills and the posterior fourth of the laterals black. Length of male sent me by the Rev. H. B. Tristram, seven inches; female six inches and a quarter.

Cabanis has separated this bird, with some others, from the genus *Saxicola,* and included it under that of *Dromolæa,* in consequence of its habits being different from those of the typical Wheatears. He has been followed by Bonaparte and others, who, by adopting, have acquiesced in the propriety of the arrangement. I have not space here to do more than account for my own reasons for preferring the retention of this species in the genus *Saxicola.* I do not for a moment question the grounds upon which this separation is made, as being contrary to the scientific rules observed by modern ornithological classifiers; but I doubt very much whether any greater degree of precision in definition is obtained, while the evil of a multiplicity of generic names, and a consequent complexity of ornithology as a science, is effected. I therefore prefer the retention of the original generic term of Bechstein, though I admit that it does not give a general idea of the structure and habits of the whole class, and therefore was perhaps originally ill-chosen. But just as *Sylviadæ,* or, as we are told it ought to be, *Sylviidæ,* is sufficient for all useful purposes as indicating a great group of birds, all of which have not sylvan habits, so I think we may accept *Saxicola* as a good generic term though all the species do not affect rocks and stones.

No better illustration could be given of the truth of

these remarks than the fact of a good and practical ornithologist like the Rev. Mr. Tristram, who adopts Cabanis's division of the genus, having the greatest possible difficulty in deciding on which side to place the Bushchat, *(Saxicola philothamna,)* which he discovered in Northern Africa, and which he has described and figured in the "Ibis," vol. i., p. 299.

The Black Wheatear is an inhabitant of the warm and southern parts of Europe, especially, being found in Spain, Sardinia, Sicily, Corsica, South of France, the Pyrenees, the Hautes and Basse Alps, the Appenines, (accidentally,) the neighbourhood of Gibraltar, and Greece. It is included in Captain Loche's list of Algerian birds. It does not appear in Mr. Carte's interesting list of the birds of the Crimea, kindly sent to me by Dr. Leith Adams.

The Rev. H. B. Tristram's account of this bird, as observed by him in Northern Africa, is so interesting that I shall transcribe his notice of it from the "Ibis," vol. i., p. 296.

"The Chats are the tribe of all others the most universally-distributed in the Desert, yet having specifically very narrow limits. They are, too, the only class of birds there which have any distinctive or conspicuous colouring. The Larks, of various species, or the Sand Grouse, may be on all sides, yet only a practised eye can detect a sign of life in the waste. But the lively Chat is seen afar; his clear bright colouring gleams in contrast with the universal brown around him. Conscious of his attractions he attempts no concealment, but relies for safety on his watchful eye and rapid movements, and, above all, on the snug retreat which he always has open before him—his hole in the rocks, or his burrow in the sand.

I think that those who are familiar with the habits of this class, will at once admit the propriety of Cabanis's separation of the genus *Dromolæa* from the old one of *Saxicola*. Strong as are the structural affinities throughout the whole, the manners of the living birds are in marked contrast. Wherever there are savage ravines, bare cliffs, reflecting a burning glare on the hungry valley, rent chasms, fearful in the unspeakable stillness which pervades the transparent atmosphere around, gorges which strike the intruder with awe, as though life, vegetable or animal, had never dared to intrude there before,—even here may a pair of Rock Chats of some species or other be detected. If a snap shot has been successful, the victim generally contrives to escape into some deep fissure to die, and frequently it is impossible to recover the spoil.

D. leucura is found only in the north of the Desert. El' Aghouat may be considered its southern limit; and it alone of the class comes up to the foot of the Atlas, on the southern slopes of which it is tolerably abundant, from Morocco to Tunis, breeding among the rocks, building a compact nest of moss and hair, and laying ordinarily four eggs, somewhat larger than those of the Wheatear, of a rich deep greenish blue, covered towards the larger end with rust-red blotches and spots. Its song is monotonous, consisting of but three notes; but the call-note is clear, loud, and musical."

The adult male has all the upper parts, except the rump, and all the lower parts, except the greater part of the tail, sooty black. Wings hair brown. Rump, upper tail coverts, and three fourths of tail below, pure white; two middle tail feathers black, the rest white, broadly barred with dark brown at the ends. Sides and beak black; tarsi glossy black.

The female only differs from the male in being smaller, and having those parts which are black in the male, of a more uniform brown, like the wings, and it is darker on the back.

The young of the year, according to Degland, resemble the female, but the brown of the middle of the abdomen is shaded with red; the wing feathers slightly fringed with grey, the primaries being terminated by a grey border, and the tail feathers with white.

My figures of this bird and its egg are taken from specimens kindly sent me by the Rev. H. B. Tristram, obtained by him during his travels in Africa.

It is also figured by Roux, Orinth. Prov., p. 197, (male;) Gould, B. of E., p. 88; Bouteille, Orinth. du Dauph, pl. 22, f. 1.

INSECTIVORÆ.

Family SAXICOLINÆ.

Genus SAXICOLA. *(Bechstein.)*

RUSSET WHEATEAR.

Saxicola stapazina.

Motacilla stapazina,	LINNÆUS; Syst., 12th. Edit., vol. i., p. 332, 1760.
" "	GMELIN, 1788.
Vitiflora rufa,	BRISSON; Orn., 1760.
" *stapazina,*	BONAPARTE, 1838.
Sylvia stapazina,	LATHAM; Ind., 1760.
" "	TEMMINCK; Man., 1st. Edit., 1815.
Saxicola stapazina,	TEMMINCK; Man., 2nd. Edit., vol. i., p. 239, 1820.
" "	CUVIER. LESSON. SCHLEGEL.
" "	KEYSERLING & BLASIUS.
" "	DEGLAND. SCHINZ.
Œnanthe stapazina,	VIEILLOT; Dict., 1818.
Muscicapa melanoleuca,	GULDENSTADT; Nov. Com. Petr., 19, p. 468.
Le cul Blanc-roux,	BUFFON; Ois., vol. v., p. 246.
Le Moteux Stapazin,	VIEILLOT; Faun. Fran., p. 189.
Traquet Stapazin,	OF THE FRENCH.
Schwarzkehliger Stein-schmäter,	OF THE GERMANS.
Monachella con la Gola Nera,	OF SAVI.

Specific Characters.—Throat more or less black; first primary shorter than the third; the two middle tail quills black, with the base white; the others white, with their extremities black. Length five inches seven lines to five inches nine lines. Length of specimen sent me by Mr. Tristram, five inches and three quarters.

THE Russet Wheatear is principally found among the rocky mountains of the south of Europe; is very common in the southern parts of Italy, in Dalmatia, in the Archipelago, and the south of France. It is also common on the rocky shores of the Mediterranean, but, according to Temminck, it is very rare in the north of Italy; seldom found in the Pyrenees, and never in the centre of Europe. It is noticed by the Hon. T. L. Powys, among the birds observed by him in the Ionian Islands. It is mentioned by Mr. Taylor among the birds found by him on the Nile; by the Rev. H. B. Tristram in Northern Africa; by Mr. Salvin in the Eastern Atlas; by Dr. Heuglin among the birds of the Red Sea, "Ibis," vol. i.; and by Mr. Carte among the birds of passage in the southern parts of the Crimea. It is also stated by Captain Loche to be found in the three provinces of Algeria. It does not occur in Mr. Hodgson's "Catalogue of Indian Birds," though the family is well represented. Neither is it included in Dr. Adams's "Birds of Cashmere." In North America the *Saxicolinæ* are represented by the genus *Sialia* of Swainson, containing however only three species.

The habits of the Russet Wheatear are very similar to those of the other members of the family, between some of which there is a very strong affinity—quite sufficient, according to Mr. Tristram's observations, to justify the inference that they may be considered as races or permanent varieties of the same species. Between *S. stapazina* and *S. homochroa*, says this gentleman, widely as they appear separated, a well-chosen series of the numerous African species of the class, "will exhibit a range of transitions so imperceptible, that it will be found very difficult without careful comparison to draw a line between one species and another."—"Ibis," vol. i., p. 432.

This bird, says Mr. Salvin, at page 307 of the same work, is found in similar situations, and appears equally distributed with *S. aurita,* whose favourite resort is among stony ground at the foot of hills or old ruins. "No difference is noticeable between the eggs of these species."

In general distribution of colour Mr. Tristram says there is much resemblance to the Desert Chat, but the bill and tarsi are one third less in length; the black of the throat does not extend so far, and in the latter, the head and back are of a more sombre isabel colour. Mr. Tristram says "that without exception the upper plumage of every bird, whether Lark, Chat Sylvian, or Sand Grouse, and also the fur of all the small mammals, and the skins of all the snakes and lizards, are of one uniform isabelline or sand-colour."— "Ibis," vol. i., p. 430.

This beautiful adaptation of colour, so important as a protection against their enemies, is, in my opinion, produced in these animals principally by means of the food. When colour is owing to the deposit of pigment, it is clear that this pigment must exist in the organic productions by which the animals are surrounded, *for it produces the same colour in them.* In the humming bird which feeds on the nectar of flowers, which being hidden, requires no colour-pigment, the hue of the plumage is owing to a peculiar sculpturing of the ultimate ramule of the colourless feather.

About March, says Count Mühle, after every fresh storm, bands of new arrivals of *S. stapazina* may be observed in Greece. They soon scatter themselves among the rocky hills, where they move about restlessly among *Emberiza cæsia*, *Surnia noctua*, and *Turdus cyanus*. They always seem angry without there being any cause of alarm, and are constantly snapping and pecking one another, although they live at peace with other birds.

They are very shy and circumspect, and build their nests in the holes of rocks, singly. The nest is made of blades of grass and the down of grass flowers, and generally contains five eggs, sea-green, sprinkled sparingly with pale-coloured spots.

Of twenty-seven eggs examined by Moquin-Tandon, from the neighbourhood of Gignac, twenty were of a uniform blue, rather darker than the eggs of the Common Wheatear; six had points, almost imperceptible, of brownish, particularly at the larger end; one had a deeper colour, with five or six spots of brown black on the greater end.

The adult male in breeding-plumage has the top of the head, nape, and upper part of back, rich buff; lower part of the back white, mottled with black; rump, upper tail coverts, and three parts of the tail beneath, white; throat, and underneath eyes and ears, upper wing coverts, and two medium tail feathers, glossy black. Wings blackish brown; secondaries fringed with grey, and the primaries underneath blackish brown; chest, abdomen, flanks, and under tail coverts, light buff, more or less deep on the chest; forehead, and a line between the black of the throat and the neck, creamy white. Beak and feet, black; iris dark brown.

In autumn, according to Degland, the top of the

head and neck, and upper part of back, are russet, shaded off into grey on the head; the chest is a brighter russet, passing into whitish on the abdomen; below the eyes, throat, and wings, black, with the feathers bordered more or less with russet, particularly the wing coverts; rump and tail as in breeding season, but with a slight border of greyish or russet at the extremity of the primaries.

The female in the breeding plumage has the head russet brown; nape and back dirty russet; throat blackish; abdomen, and a large band over the eye, whitish; scapularies black, terminated with russet; wings blackish brown, the quills bordered with russet; tail as in the male, but with the black more extended.

My figures of this bird and its egg are from specimens kindly sent to me by Mr. Tristram.

Figured by Edwards, plate 31; Guldenstadt, Nov. Comm., petr. xix., p. 468, tab. xv.; Naumann, taf. 90, figs. 1 and 2; Roux. Ornith. Prov., pl. 199, (male,) fig. 1, (female,) fig. 2; Bouteille, Ornith. du Dauph., pl. 22, fig. 2; Gould, B. of E., pl. 91.

INSECTIVORÆ.
Family SAXICOLINÆ.
Genus SAXICOLA. *(Bechstein.)*

BLACK-EARED WHEATEAR.

Saxicola aurita.

Vitiflora rubescens,	BRISSON; Orn., 1760.
" *aurita,*	BONAPARTE, 1838.
Motacilla stapazina, (var. B.,)	GMELIN; Syst., vol. i., p. 966, 1788.
Sylvia stapazina, (var. B.,)	LATHAM; Ind., p. 531, 1790.
Saxicola aurita,	TEMMINCK; Man., 2nd. Edit., vol. i., p. 241, 1820.
" "	SCHINZ, 1840. SCHLEGEL, 1844.
Œnanthe albicollis,	VIEILLOT; Faun. Fr., p. 190.
Le Moteux Regnauby,	VIEILLOT.
Traquet Oreillard,	OF THE FRENCH.
Schwarzöhriger Steinschmätzer,	OF THE GERMANS.
Monachella,	OF SAVI AND STORR.

Specific Characters.—Throat always white or whitish; rest as in *Stapazina*. Length five inches seven to nine lines.—TEMMINCK. Length of specimen sent me by Mr. Tristram, six inches.

THE Black-eared Wheatear is an inhabitant of the middle-sized mountains of the south of Europe, and, according to Temminck, is more common in the north of Italy than the preceding species. It is frequently found

on the shores of the Mediterranean, in the Appenines, in the Illyrian Provinces, in Sardinia, and Neapolitan States, but, like the preceding species, never in the centre of Europe. It appears in the south of France in spring, but never in great numbers. It is mentioned by Count Mühle as being found in Greece. The Hon. T. L. Powys, in his very interesting description of the birds of the Ionian Islands, ("Ibis," April, 1860, page 140,) says that this bird is the least common of the three species observed there. It arrives with the Common and Russet Wheatear, in March. It is included in Captain Loche's list of Algerian birds; in that of Mr. Tristram in Northern Africa; and Mr. Salvin in the Eastern Atlas. Schlegel gives Arabia and Egypt as localities. It does not appear to occur in India or the New World.

The habits of the Eared Wheatear are very similar to those of *S. stapazina*. It is found in nearly the same localities, and was long confounded with it. They are found in the most retired and arid regions, and together imitate the songs of other birds. The two species, according to Degland, are to be distinguished by the throat being at all times white in *aurita*, and always more or less black in *stapazina*. The tarsi also are shorter, and the colour of the eyes more lively than those of the latter bird. On this subject we have also the high testimony of Count Mühle, who remarks, that although the identity of the two species has been advocated by Bonelli, Calvi, and Ritter Von Marmora, he has had abundant opportunity in Greece of satisfying himself that they are distinct. In the specimens which are now before me the buff colour is much lighter in *aurita* than *stapazina*, and the tarsi are slightly shorter; the curve at the end of the upper mandible is also shorter. Altogether it is very difficult to come to any other conclusion than that they are races of the same bird. On the whole, however, it appears to have the requisites of modern ornithology to constitute a species.

The Rev. H. B. Tristram, whose knowledge of these birds is very considerable, and derived from that sure source, practical acquaintance in their own homes, writes me word—"*S. stapazina* and *S. aurita* differ from the youngest to the oldest stage constantly. They are both very common on the sides of the Atlas, especially the southern. I have taken many nests, but never knew the two cross. There is not the slightest difference in the sexes. The nests are alike. In the eggs, to judge from a small induction, *stapazina* has more spots collected in a zone at the broad end; in *aurita* they are more generally diffused over the surface. They build in niches in rocks and ruins, and on the ground or steep banks, or among the stumps of old trees. They differ among themselves, I should say, as Whinchat and Stonechat, and yet it is very curious how close, and yet how distinct they are. I have got a nest of each, taken from the same ruins."

On the other hand, Moquin-Tandon ("Revue et Mag. de Zoologie," July, 1858,) says,—"I for a long time thought the eggs of *S. stapazina* were unicolorous, and those of *S. aurita* marked by brownish spots on the greater end. Fresh observations have shewn me that the eggs of the two species are exactly alike, which has given additional weight to the opinions of Bonelli and Prince Charles Bonaparte, that these two *Saxicolæ* are one and the same species."

"The favourite resort of the Eared Wheatear," says Mr. Salvin, ("Ibis," vol. i., page 307,) "is among stony ground at the foot of the hills; and in such places it may be looked for and generally found. Roman ruins also are much frequented. We obtained two nests from the Madracen, where they were placed in the interstices of the stone of that building. Usually the nests were close by or under a large fragment of rock."

Like *stapazina*, the Eared Wheatear builds its nest among rocks and stones near the ground. The nest is deep and wide, and is not made with much care. It is formed of dry grasses, wool, hair, etc., in which is deposited five or six eggs, of a greenish blue, generally deeper coloured than those of *stapazina*, and with the spots thicker, and more coloured with brown or rust red.

The male in breeding plumage has the head, nape and back, of a light buff. Rump, throat, and two-thirds of tail below, white; abdomen and under tail coverts, creamy white, more or less shaded with light buff; scapularies a mixture of buff and black feathers; greater wing coverts, two upper tail feathers, lower third of tail underneath, and a band extending from the gape along the entire cheek and side of head, glossy black. Wing primaries hair brown, secondaries darker; beak and legs, black; iris, dark brown.

The female, according to Degland, differs sensibly from the male during the breeding season; the side of the head is brown, mixed with russet; the throat is dirty white; the wings less black, and that of the tail quills less extended. In autumn the changes of both sexes are similar: more russet on the upper and lower parts, and the feathers of the wings deeply bordered with russet.

The young before the first moult resemble the young of the Rock Thrush: an ashy russet, darker below, with each feather bordered with brown, and marked in the centre with a yellowish spot; middle of the abdomen and under tail coverts, of this tint; middle and greater wing coverts broadly bordered with russet. After the first moult the young only differ from the female in autumn, by the wing having no trace of brown on the side of the head, and by the throat being russet.

My figures of this bird and its egg are from specimens with which I have been obligingly favoured by the Rev. H. B. Tristram.

Figured by Brisson, Orn., vol. iii., pl. 25, fig. 4; Edwards, pl. 31, (a good figure, but given as the female of *S. stapazina;*) Vieillot, Faun. Fr., pl. 85, figs. 1, 2, and 3; Roux, Ornith. Prov., pl. 200, (adult male;) Bouteille, Ornith. du Dauph., pl. 22, fig. 4; Gould, B. of E., pl. 92.

INSECTIVORÆ.
Family SAXICOLINÆ.
Genus SAXICOLA. *(Bechstein.)*

PIED WHEATEAR.

Saxicola leucomela.

Motacilla leucomela,	PALLAS; Nov. Comm. Petrop., xiv., p. 584.
Saxicola leucomela,	TEMMINCK; Man., 2nd. Edit., vol. iii., p. 166.
" "	KEYSERLING AND BLASIUS.
" "	SCHINZ. SCHLEGEL.
" "	DEGLAND.
" *lugens,*	LICHTENSTEIN; Cat. des doubles du Mus de Berlin, p. 33.
Vitiflora leucomela,	BONAPARTE.
Traquet leucomèle, Motteux leucomèle, Motteux pleschanka,	OF THE FRENCH.
Elster Steinschmätzer,	OF THE GERMANS.

Specific Characters.—First primary shorter than the third; basal half of two middle tail quills white, the other half black; upper tail coverts, white; under tail coverts slightly tinged with russet. Length of male specimen sent me by Mr. Tristram, six inches; carpus to tip three inches and a half; tarsus one inch.

This bird is an inhabitant of the north of Europe, especially the southern parts of Russia, the Daouria, the Altai Mountains, and Lapland. There is a species found in India, which only differs from the present in having the under tail coverts white instead of pale russet. This is, however, thought sufficient, by Gray and others, to separate it from the *leucomela* of Temminck; the consequence of which is, we have two *leucomelas*, that of Latham in India, and that of Temminck in Europe.

Dr. Leith Adams informs me that the bird mentioned in Mr. Carte's "Observations on the Climate and Zoology of the Crimea," as being observed there during the passage in April, is the Indian species, with the white under tail coverts. This makes it extremely probable that there is only one species. Temminck mentions the Levant and Crimea as localities for his species. The European bird also differs in the more or less deep shade of the russet colour of the under tail coverts. One variety has been called and described as a different species under the name of *S. lugens* by Lichtenstein. This has also been found in Greece by Count Mühle. It inhabits the Levant, Egypt, and Nubia; while Temminck's typical species is found more especially in the Ural and Siberia. All modern authors, however, agree in considering that the *darker* under tail coverts of *S. lugens* do not entitle it to specific distinction from *S. leucomela*. Why then should a *lighter* colouring of the same feathers, and part of the shafts of primaries, be adduced as evidence of a specific difference between the Indian and European *leucomela?*

Mr. Tristram writes me word that he doubts the identity of the African and Indian *leucomela* with *lugens;* but he adds, "take an Egyptian or Arabian

bird, and you will find an intermediate gradation. Are they not all races of the same species?"

The Pied Wheatear has habits very similar to its congeners. It builds in the clefts of rocks, sometimes among heaps of stones. Temminck says it also builds in the banks of rivers, in the holes made by wasps. It lays four or five eggs, having the general character of those of the family. Its food is beetles and other insects.

The adult male in breeding plumage has the summit of the head, nape, rump, lower part of chest, and abdomen, pure white; side of the head, throat, front of neck, space between the eyes and beak, and greater wing coverts, sooty black. Wings brown, the secondaries slightly bordered with white. Tail white for two thirds of its length, the end and two middle tail feathers black; under tail coverts light russet; beak, feet, and iris, black.

The female is ashy brown above, with a paler tint on the head; ashy below, with the eyebrows and throat white.

Birds of the year, according to Degland, have the head varied with white and brown; feathers of the back and wing coverts bordered with russet; throat and front of neck barred with russet and black; abdomen dirty white. The young males have the flanks ashy grey.

Figured also by Pallas, Nov. Comm. Petr., 14, pl. 22, fig. 3; Temminck, pl. color. 257, f. 3, adult male; Guldenstadt, Nov. Comm. Petrop., vol. xix., p. 468, pl. 15.

INSECTIVORÆ.
Family *SAXICOLINÆ.*
Genus Saxicola. *(Bechstein.)*

MENETRIES' WHEATEAR.

Saxicola saltatrix.

Saxicola saltator,	Menetries; Cat. Cauc., 1836.
" *saltatrix,*	Keyserling and Blasius; Die Wirbelt, 1840.
" "	Schlegel; Revue, 1844.
" "	Degland, 1849.
Vitiflora saltatrix,	Bonaparte, 1838.
Traquet Oriental,	Of the French.
Ostlicher Steinschmätzer,	Of the Germans.

Specific Characters.—The first primary equal in length to the fourth feather of the greater wing coverts; the second a little longer than the fifth, the second and fourth about equal, but shorter than the third, which is the longest in the wing. Length of a specimen sent me by Mr. Tristram, and which is figured, six inches and three quarters; carpus to tip of wing four inches; tail two inches.

This bird is an inhabitant of Egypt, Nubia, the Ural Mountains, Greece, and borders of the Caspian Sea. It is mentioned by Mr. Tristram, in his list of the Birds of Palestine, and by Dr. Hueglin, among those collected by him during a voyage in the Red Sea, ("Ibis," vol. i., pages 29 and 341.) Dr. Hueglin

found it in the Danakil country, between the Peninsula of Buri and the Gulf of Tadjura, on the Somali coast, and in Southern Arabia.

There is nothing to add about the habits of this bird, which Mr. Tristram says are precisely similar to those of the Common Wheatear.

Head, nape, and back down to the rump, and wing coverts, a mixture of buff with olivaceous green; rump and basal half of tail pure white. Wings brown, the primaries after the fourth lightly, and the secondaries more deeply bordered with the same colour as the back, but brighter, the same tint as the fringe being shewn where the colour of the back passes into the white of the rump. As usual in all the Wheatears I have described, the two middle tail feathers are black after the first third from base. Throat, neck, and lower part of abdomen, dirty white; the chest and upper part of the abdomen, under wing and tail coverts, light buff.

The colours above described of the feathers of the body, are produced by tints at the *extremities* of the feathers only. The real colour of all the feathers below the surface is black, like that of *leucomela*, a point well worthy of attention in looking at the affinities of these birds, and estimating how far the variations in colour may be owing to climatic causes and mode of living.

The feathers covering the ears are a darker buff, with a light line extending over them from the angle of the eye. Feet black, the hinder claw more strongly curved than the anterior ones; beak horn-coloured.

I am indebted for the bird from which my drawing is taken, as well as those of bird and egg of the last species, to the kindness of the Rev. H. B. Tristram. My figure of the bird is from a male specimen, killed in Egypt, February 5th., 1852.

INSECTIVORÆ.
Family MOTACILLIDÆ.
Genus Motacilla. *(Latham.)*

Generic Characters.—Beak slender, straight, awl-shaped, cylindrical, and angulated between the nostrils; the edge of the inferior mandible compressed inwards. Nostrils basal, lateral-ovoid, partly closed by a naked membrane. Tarsi as long again as the middle toe; three toes in front and one behind, the outer toe of the three joined to the middle toe at its base; the claw of the hinder toe longer than those in front, which are very small. Tail very long and nearly even at the end, having twelve feathers. Wings of moderate size, first quill the longest, second and third equal, and nearly as long as the first. The tertials very long.

YELLOW-HEADED WAGTAIL.

Motacilla citreola.

Motacilla citreola,		Pallas; Voy., vol. iii., p. 696, 1776.
"	"	Gmelin; Syst., i., p. 962, 1788.
"	"	Latham; Ind., 1790.
"	"	Temminck; Man., p. 259, 1820.
"	"	Keyserling and Blasius; Die Wirbelt, 1840.
"	"	Schinz; Europ. Faun., 1840.
"	"	Schlegel, 1844. Degland, 1849.
"	"	Middendorff; Sibirische Reise, ii., p. 163.
Budytes citreola,		Bonaparte, 1838.

Bergeronnette citrine,	OF THE FRENCH.
Citronengelbe Schafstelze,	OF THE GERMANS.
Yellow-headed Wagtail,	PENNANT; Arct. Zoology.

Specific Characters.—Beak and scapularies bluish grey; head, neck, throat, and all inferior parts of the body, citron yellow; the lateral tail feathers pure white. In the female top of the head and cheeks ashy grey. Length six and a half to seven inches.

THE Wagtails form a group of birds always interesting to the naturalist. They are among the most beautiful and elegant of the feathered tribes, and there are few people who have not watched their graceful movements among our rocky streams without pleasure.

They are also interesting studies to the philosophic naturalist—for they present him with some pleasing problems as to the distinction between species and variety and race. "Natural selection" has been busy with the group, and without however shewing any tendency to develop a Pelican or a *Balæniceps rex* out of the delicate Wagtail, it has given to one a darker head, and to another a gayer coat, which I doubt not will, in that extensive future which we are told to expect, have their due influence over the deluded eyes of the weaker Wagtail sex.

There are eight European Wagtails described by authors, five of which are observed in England. Degland however has reduced this eight to four. He leaves out *M. lugubris*, Pallas, as of uncertain occurrence in Europe, and he considers *M. yarrelli*, with which it is thought identical by authors, as a variety of our White Wagtail, the *Motacilla alba* of Linnæus.

M. cinereo-capilla of Savi, *M. melanocephala* and

M. flaveola, (Ray's Wagtail,) are considered by both Degland and Schlegel, to be races or varieties of *M. flava*, (our Grey-headed Wagtail.)

Mr. Tristram writes me word that he cannot satisfy himself of the specific distinction of *M. flava* from *M. cinereocephala*, and that he can shew every intermediate gradation between *M. flaveola* and *M. melanocephala*.

This subject is very well treated by Dr. Zander in "Naumannia," 1858, Part 3, p. 239. Dr. Zander considers that all European Wagtails are varieties of *M. alba*, *M. boarula*, *M. citreola*, and *M. flava*, Linnæus. He says he considers much of the difficulty arises from the fact that the intermediate varieties are not so frequently seen as the so-called species. He describes how perplexing are the changes produced by a substitution of black or yellow for grey, or by the passing of grey into grey yellow. "The grey goes through all shades till it comes to the clearest black, and the eye stripe becomes less, until hardly seen." He also thinks that the various colours in the females and young are not good specific indications. The clear black head predominates in warm, and the black-grey head in temperate climates, the grey-yellow head being peculiar to England.

Under these circumstances, and after consulting various specimens, and leaving the English species for others to discuss, I shall introduce into this work the subject of the present notice as distinct; and *M. flava-cinereocephala*, and *M. flava-melanocephala*, as probable permanent varieties or races. And I do this the more willingly because I think the more doubtful a species, the greater the necessity for making it thoroughly well known.

The Yellow-headed Wagtail is an inhabitant of

Eastern Russia, Siberia, and Bokhara. It is rarely found more south, though Calvi has introduced it among the birds of Liguria, and Temminck has given the Crimea, Hungary, and the Archipelago, as probable localities. It is therefore not much known to naturalists, and we must accompany Pennant, Pallas, or Middendorff, into the far cold arctic regions, to gain a glimpse of its whereabouts.

"This species," says Middendorff, "breeds very rarely in Boganída, (71° N., Br.) In the S. E. I missed it entirely. Gould falls into the same mistake as Pallas in stating that the European Yellow Wagtail exceeds *M. citreola* in size. It is just the contrary. The colour of the back of my Siberian species is also blackish, with a lead-coloured tint, and not greenish, as it is represented by Gould. The summer dress of the old female seems hitherto to have escaped observation; what Pallas says respecting it is indefiinte, and Gould's drawing does not agree with the Siberian form. The female in summer dress has the top of the head lead-grey, with a greenish tint, and the yellow of the throat is separated from the yellow stripe over the eyes by a grey band."

Of its habits and propagation nothing is known, as Middendorff does not mention the subject further than in the passage quoted.

The adult male has the top of the head, cheeks, and inferior parts generally vivid and pure citron yellow; on the occiput is a large black band, in the form of a cross; nape, back, crop, and flanks, lead-colour; middle and great wing coverts bordered and terminated with pure white; primaries and rectrices, blackish, except the two lateral tail feathers, which are pure white; feet and legs brown; posterior claw longer than the toe. The males and females have no black cross on the occiput after the autumn moult.

The distinctive characters of the female have been already noticed, and will be further observed in my figure, which is an adult female in summer plumage, after Middendorff.

Figured by Gould, pl. 144; Middendorff, Sibirische Reise, ii., pl. 14, fig. 4.

INSECTIVORÆ.
Family *MOTACILLIDÆ.*
Genus MOTACILLA. *(Latham.)*

GREY-HEADED YELLOW WAGTAIL.

Motacilla flava, (Lin. nec Ray.) Var. *cinereo-capilla.*

Motacilla cinereo-capilla,		Savi; Orn. Tosc. iii, p. 216.
"	" "	Bonaparte; Icon. d. Faun. Ital., pl. 31, f. 2.
"	" "	Degland. Dubois.
"	" "	Cabanis.
"	" "	Selys-Longchamps.
"	*flava cinereo-capilla,*	Schlegel; Revue.
"	*feldeggi,*	Michaellis.
"	*dalmatica,*	Bruch.
Bergeronnette de printems a tête grise,		Of the French.
Grauköpfige gelbe Schafstelze,		Of the Germans.

Specific Characters.—Feathers of the back and rump olive green; throat white. No white or yellow stripe over the eye. All the rest as in *M. flava.* Length about six inches.

Nothing can better illustrate the difficulties which attend the determination of species when slight alterations in colour are taken as sufficiently distinctive, than this and the following bird. As Dr. Zander has well remarked, we find the two ends of the series of varieties, and constitute them species. The intermediate forms do not come under our observation so frequently, and we therefore lose the significance of the serial affinities. Believing, as I do, that much of the system of determining species in Natural History in modern days is deficient in sound scientific principle, I have no occasion to seek for a solution of the difficulty in the theory of transmutation. I think that differences of climate and food are all-sufficient to produce a great majority of the variations we meet with; and as it is more than probable that the world contains a vast number of special cases wherein these influences of food and climate operate distinctly, I have no difficulty in accounting for the variation of species, or of satisfying myself that the difference of a feather here or there is not sufficient to justify the splitting up of our naturally defined genera and families into an interminable long list of Greek derivatives, quite sufficient to frighten away nine tenths of the students of nature from the most beautiful and instructive of all pursuits.

The Grey-headed Yellow Wagtail is a common bird during the summer in Italy and other hot parts of Europe, as Dalmatia, Tuscany, Sicily, Sardinia, and the south of France and Spain. This is its area, but it is also occasionally found in Belgium, and it has been taken by M. De Selys-Longchamps in the neighbourhood of Liege. It has also been found in Nubia and Egypt.

Of its habits we have the following in the "Oiseaux de la Belgique," of M. Ch. F. Dubois:—

"The extensive plains, meadows, and marshes are the places which these birds more especially frequent; they like a moist soil among osiers and reeds, and other aquatic plants, especially the male, who in the breeding season will remain for hours in these retreats appealing to his mate with his soft and tender notes. There is much agility in all their movements, but they are excessively timid, and are driven away by the least noise. Notwithstanding they are taken in numbers in Italy, where the pleasures of the table supersede any regard for the utility of these birds. The nest is placed at a slight elevation on the ground, either among the grass or corn, and is formed of roots, blades of grass, and moss; the interior is made soft and warm by wool. The eggs are four or five in number, and do not sensibly differ from those of the Yellow Wagtail."

The "specific characters" by which this notice is headed, are those by which this variety may generally be distinguished, but they are by no means constant. Mr. Tristram has sent me two specimens of *M. flava* of Linnæus, (known to British naturalists by the name *M. neglecta*, adopted by Gould and Yarrell,) shot in Algeria, one in May, 1856, the other in May, 1857. In the former the head is grey; the white mark extends from the posterior angle of the eye for five lines and a half, but not over it at all, and the throat is pure white. In the latter, a male, the white stripe extends as usual over the eye, but the throat, like the rest of the body, is yellow, with this exception, *that the feathers at the base of the under mandible, and a line separating the grey of the head from the yellow of the throat, are white.* These are evidently transitional varieties, and did they point to real structural changes, would be valuable supports of the natural selection theory.

Of the adult male the top of the head, the nape, and cheeks are of a grey lead-colour; the back and rump bright olive green; throat, sides, and front of the neck pure white; rest of the under parts bright and deep yellow, having the flanks more or less olive green; wing coverts bordered with olive green. Wings and tail as in the Yellow Wagtail; beak, iris, and feet, brown.

In the female the olive green of the upper parts has a tinge of russet on the head and back.

In the young the upper parts, according to Degland, are ashy green, with the head olive-coloured, yellow below, with the throat white, and a *yellow superciliary edge.*

It is figured by Bonaparte in "Fauna Italica," pl. 31, f. 2, and by Dubois, "Oiseaux de la Belgique," part 77, pl. 93, male and female.

INSECTIVORÆ.
Family MOTACILLIDÆ.
Genus MOTACILLA. (Latham.)

BLACK-HEADED YELLOW WAGTAIL.

Motacilla flava, (Lin.,) Var. *melanocephala.*

Motacilla melanocephala,	BONAPARTE; Icon. d. Faun. Ital. pl. 31, fig. 3.
" "	DEGLAND. LICHTENSTEIN.
" "	DUBOIS.
" *flava-melanocephala,*	SCHLEGEL; Revue.
" *viridis,*	GMELIN. SCOPOLI. HODGSON.
Budytes melanocephala,	MENETRIES.
Bergeronette, or Hochqueue a tête Noire,	OF THE FRENCH.
Schwarzköpfige gelbe schafstelze,	OF THE GERMANS.

Specific Characters.—Top of the head and cheeks a deep black; beak black; throat white. Length about six inches.

THIS so-called species was described and figured by Prince Charles Bonaparte, in the "Fauna Italica," pl. 31. Its claims to specific distinction have, however, been doubted by Temminck and others. Mr. Tristram writes me word that he has seen every gradation of colour from *M. flaveola*, (our Yellow Wagtail,) up to the present bird, and he has sent me specimens which bear out this opinion. It is, in fact, very probable that all the European Yellow Wagtails are permanent varieties or races of the same type. Mr. Tristram writes, "I have obtained *M. flaveola*, (Ray's Wagtail,) in Morocco, where it is certainly the general, if not the only variety. In Spain and Portugal I have got it, and not *M. flava*, and I presume that on the Atlantic coast *M. flaveola* is the commonest form. In Algeria *M. flava* is almost universal, but I have twice shot *M. flaveola* in its passage in the Bay of Algiers. At Tunis, and further east, I have only found *M. flava* and *M. cinereo-capilla*. At Pyles (Navarino,) I got *M. melanocephala*, which is also very common at Athens, where I saw no other. In Palestine I got *M. melanocephala* only, but I saw at Jerusalem, in Dr. Rotte's room, a skin of *M. flava*. Thus it appears to me that the varieties glide into one another, the black on the head increasing in intensity as we go eastwards."

These are very interesting remarks, and clearly lean towards the very strong inference that these birds have a common origin. Count Mühle has taken the opposite view, and gives the following as his reasons. He says in the first place, that in Greece the plumage of *M. flava* is the same as with us, that it never mixes with *melanocephala*, and that, while the former is found in the districts of Lavadien, Malo, and Lamia, the latter occurs in the Morea; and that, where *M. flava* is found, there also shall we meet with *M. alba*, but never *M. melanocephala*. He further remarks that *M. melanocephala* goes away early with *Merops apiaster* and *Emberiza cæsia*, while *M. flava* may be found in winter, and that among hundreds of specimens of

M. melanocephala, which he had seen, none were in a transition state. M. Dubois endorses Count Mühle's views, and gives figures and description of the bird.

Dr. Leith Adams informs me that this bird is identical with the *M. viridis* of Scopoli, and that consequently this name ought to have priority. Dr. Adams remarks: —"*B. viridis* and *melanocephala* are identical, and I have Mr. Blyth's authority for this opinion, (Cat. Mus. As. Soc., No. 775-776, and Append., p. 325,) who remarks, 'the birds acquire the blue-grey feathers on the head at the vernal moult, which change oftentimes to black.' It is found in the Punjaub, Scinde, and India generally, as well as in North Africa."

With regard to the priority of name, as I have only noticed this bird as a variety of *M. flava,* any alteration would be unnecessary. The above remarks and quotation by Dr. Adams, give further strength to the view I have taken of the want of specific distinction in this bird.

The Black-headed Yellow Wagtail, or, as it is called by Dubois, the Blackcap Wagtail, is found generally, as it has appeared from the above remarks, in Algeria, Egypt, Nubia, Arabia, Syria, Persia, and in Bokhara. According to Mühle it is common in the Morea. It occurs in Dalmatia, Sicily, and the Caucasus. In Germany, France, and Belgium it is accidental, and rare in Italy. Taken by W. H. Simpson, Esq., at Missolonghi, in Greece, June, 1859.

In its habits it varies but little from the variety last described. It is fond of pastures, plains, and marshes. Dubois says that it is constantly on the ground; but that it also is frequently found perching on the branches of bushes, and on the stems of willows and osiers; that they are very lively in their movements, but he deprives them of their usual gentle character by declaring that they are *"farouches et sauvages."*

The same author tells us they feed on flies, gnats, moths, and beetles, both in the perfect and larva state; that they nest on the ground, or in a bush on the borders of fields and plains. The nest is made of blades of dry grass, small roots, and moss, lined inside with fine grass or wool. They lay from four to six eggs.

The adult in breeding plumage has the cheeks, top of the head, nape, and upper tail feathers deep black; back olive green, but not so dark as in the preceding variety; a beautiful yellow, or, as Degland has well expressed it, *d'un beau jaune jonquille,* below; the crop and part of the flanks more or less shaded with olive green. Wing coverts olive green, bordered with grey; primaries and secondaries hair brown, the latter broadly bordered with grey, outer tail feathers white, the inner web black at basal half; beak, feet, and iris, brown.

In the young, according to Degland, the upper parts are olive grey, with the nape ashy and the head blackish, darker in front, and above the eyes and ears; yellowish below, with the throat whitish.

M. feldeggii is a transitional variety between this and the last noticed.

My figures of the bird and its egg are taken from specimens kindly sent me by the Rev. H. B. Tristram.

It is also figured by Bonaparte, in Fauna Italica, pl. 31, fig. 2; Gould, B. of E.; Rüppell, Atlas Reise Afric, pl. 33; Dubois, Oiseaux de la Belgique, pt. 48, pl. 94, male and female.

INSECTIVORÆ.

Family *MOTACILLIDÆ.*

Genus MOTACILLA. *(Latham.)*

SOMBRE WAGTAIL.

Motacilla lugubris.

Motacilla lugubris,		PALLAS; Fauna Rossica.
"	"	SIEBALD; Fauna Japan., Tab. xxv.
"	"	KITTLITZ; Taf. 21, f. i.
"	"	BONAPARTE; Compar. list, 1838.
"	"	GOULD; B. of E., pl. 142.
"	"	TEMMINCK; Orn. Man., 1836, (but not of 1820.)
"	*leucoptera,*	VIGORS.
"	*lugens,*	ILLIGER.
"	"	BONAPARTE; Conspectus, p. 250, 1850.
"	*alba,* var. *lugens,*	MIDDENDORFF; Theil. ii., p. 166.
Schwarze Bachstelze,		MEYER; Orn. Tasch.

Specific Characters.—Head, nape, beak, scapularies, tertials, upper tail coverts, and upper tail feathers sooty black *in winter plumage.* Rump ash grey; basal half of primaries, except first, and the secondaries pure white at all seasons.

Length of male sent me by Mr. Tristram, seven inches and two fifths; of female six inches and seven tenths. From carpal joint to tip three inches and a half; tail three inches and three fifths; beak from gape seven tenths of an inch. Breadth of lower mandible at gape one fifth of an inch; tarsus one inch.

This very distinct and striking species was excluded by Schlegel and Bonaparte from the European list, the former stating we had no proof of its existence in Europe. The Rev. H. B. Tristram has however been kind enough to draw my attention to some recent captures in Turkey and the Crimea, which have confirmed the original notice of this bird by Pallas, on the borders of the Black Sea.

This bird is very distinct from either *M. yarrellii* or *M. alba*, and may probably be considered typical of the pied races.

I have been favoured with the following notes upon this species, by Mr. Tristram:—"The bird figured by Roux, Orn. Prov., under this name, and also that described by Temminck, in 1820, is merely the *Motacilla yarrellii* of Gould. Though Temminck corrected this error in his edition of 1836, and suppressed all that he had formerly written on the subject, yet these authors have been implicitly followed in their mistakes by almost all subsequent writers. So much easier is it to perpetuate error than to correct it.

Bonaparte, who had in his catalogue included *M. lugubris* among the Birds of Europe, in his later work, the "Conspectus," while acknowledging the specific value of *M. lugubris*, excludes it from the Birds of Europe, having only seen Japanese specimens. Pallas, however, found it on the shores of the Black Sea, and it has since been frequently obtained in Turkey. Several specimens were sent home by officers engaged in the Crimean war, which had been obtained near Sebastopol, some of which I have had the pleasure of examining. It winters regularly in Egypt and Nubia, which appear to be its western limits, and where it meets the *M. alba* of Europe. Thus we find one form, *M.*

alba, with its western variety, *M. yarrellii*, extending over the whole of Europe and North Africa, and another, *M. lugubris*, occupying the vast continent of Asia, and in its western limits disputing its territory with *M. alba*. It would also appear from the remarks of Middendorff, that as *M. alba* varies in its western habitat from the typical form, so does *M. lugubris* in the extreme east become more marked in its coloration.

In all stages however it may at once be distinguished from any variety of *M. alba*, by the wing primaries, more than half of the upper portion of which are pure white, while a white fringe, broader in summer than in winter, runs along the outer edge of the secondaries. The middle wing coverts are also pure white."

The adult male in winter plumage has the whole of the upper parts, except the rump, black, owing to the ends of the feathers being of that colour; the basal half of the feather is ash grey. Upper tail feathers, black; rump ash grey, mingled more or less with dusky feathers. Throat, lore, ear coverts, belly, under tail and wing coverts, and two outer tail feathers on each side, white. The white of the belly rather tinged with cream-colour, and the second tail feathers on each side having a slight border of blackish brown. First four wing primaries of nearly equal length, the second longest, the fifth, sixth, and seventh each one third of an inch shorter than the preceding feather; seventh and eighth of equal length. First primary entirely black, all the rest and the secondaries having the basal half of each feather pure white; tips of the inner five primaries, and a border on the inner web of the secondaries, white. Lesser wing coverts and distal half of primaries black; distal half of secondaries black, except the outer border; middle wing coverts with the inner web black, and the outer white, the black prevailing more as they approach the entirely black lesser coverts. Beak, feet, and tarsi black.

The female only differs from the male in being half an inch less, in having less white about the wings, and that of the belly and throat more decidedly cream-colour.

The specimen figured is a male sent me by Mr. Tristram, and marked "Assouan, Feb. 2, 1860, W. C. P. Medlycott."

It has also been figured by Gould, pl. 142.

INSECTIVORÆ.
Family *MOTACILLIDÆ.*
Genus ANTHUS. *(Bechstein.)*

Generic Characters.—Beak straight, slender, cylindrical, and compressed towards the point, with the edges bent inwards near the middle; base of the upper mandible rather elevated, and the tips slightly hollowed out. Nostrils basal, lateral, partly covered with a membrane; tarsi elongated; middle and external toes united at their base; posterior toe very long, and the claw more or less curved; the longest wing tertial as long, or longer than the longest primary. Tail composed of twelve quills, and emarginated.

RED-THROATED PIPIT.

Anthus cervinus.

Motacilla cervina,	PALLAS; Zoog., vol. i., p. 511, (1811.)
Anthus rufogularis,	BREHM; Vogel Deut, (1831,) p. 320.
" "	TEMMINCK; Man., (1835.)
" "	CH. BONAPARTE; (1838.)
" "	SCHINZ. DUBOIS. NAUMANN.
" *cervinus,*	KEYSERLING AND BLASIUS, (1840.)
" *pratensis rufigularis,*	SCHLEGEL; Revue, 1844.
" *cecilii,*	AUDOUIN.
" *rosaceus,*	HODGSON.
Pipit a gorge Rousse,	OF THE FRENCH.
Braunkehliger Wiesenpieper,	OF THE GERMANS.

Specific Characters.—Feathers of upper parts black, distinctly bordered with grey, so as to give a splendid appearance to the plumage. Cheeks, throat, and breast of the male, and throat of the female, russet red, with longitudinal dark spots. Hind toe of equal length with the claw; the latter as much or more curved than that of the Rock Pipit.

Length of male six inches and a half; from carpus to tip three inches and a half; beak from gape three quarters of an inch; beak along ridge of upper mandible half an inch; tarsus nine tenths of an inch; hind toe two fifths of an inch; claw two fifths of an inch; middle toe seven tenths of an inch, and its claw a quarter of an inch. Length of female six inches; carpus to tip three inches and one tenth; beak from gape seven tenths of an inch; beak on upper ridge two fifths of an inch; rest as male.

THE Pipits are a very natural but distinct family, closely allied to the Wagtails on the one side, and to the true Larks on the other. They are also very similar to each other, differing principally in colour, and in the shape of the hind claw. Each species is in fact adapted to the circumstances of its existence. Our own British species illustrates this very well. The Tree Pipit living principally on trees or bushes, has the hind claw short and curved; the Meadow Pipit, which lives more on the ground, has the hind claw double the length of the former, but nearly straight; while the Rock Pipit, which lives upon insects and seeds found on the mud of rivers, has the hind claw considerably curved, *which enables it to secure a firm footing on the mud.* This bird is tolerably plentiful in the neighbourhood of Colchester, and my friend Dr. Maclean, who has studied the habits of birds for many years with great care, assures me that when disturbed on the banks of the river, it never lights upon the turf, but always upon the mud by the river side. I have myself verified this statement.

The Red-throated Pipit belongs to the Rock Pipit

branch of the family; its claw being much curved. There has been much confusion about the bird in consequence of this fact being overlooked. Schlegel, Degland, and others have considered it a local variety of *A. pratensis*. But if it is a local variety or race of anything, it must be of *A. obscurus*, (Rock Pipit,) and not of the Meadow.

The Red-throated Pipit is an inhabitant of Northern Europe and Northern Africa. Middendorff, ("Sibirische reise," vol. ii., p. 165,) remarks, "It is generally *Anthus rupestris*, (Rock Lark,) that is considered the northern representative of the genus. I have not met with one in North Siberia for years, and only *exceptionally* on the European coast of the Russian or Northern Ocean. There is in the extreme north of the old world the *A. cervinus*, Pallas, in great multitudes."

It is found plentifully in Egypt, Nubia, Greece, Turkey, and Barbary, during the winter. I have been favoured with the following very interesting account of the discovery of this bird in East Finmark, by Alfred Newton, Esq., of Elvedon, who has also most obligingly sent me the skins, from which my figures are taken:—

"On the 22nd. of June, 1855, a few days after our arrival at Wadsö, in East Finmark, Mr. W. H. Simpson and I, in the course of a birds'-nesting walk to the north-east of the town, to the distance perhaps of a couple of English miles, came upon a bog, whose appearance held out greater promise to our ornithological appetites than we had hitherto met with in Norway. We had crossed the meadows near the houses, where Temminck's Stint and the Shore Lark were thrilling out their glad notes, and traversed a low ridge of barren moor, when the solicitude of a pair of Golden Plovers plainly told us that they had eggs or young near us.

A Dunlin's nest was speedily found, and the bird procured to identify it, for we had hopes of all sorts of waders in that remote district. A little while after, as I was cautiously picking my way over the treacherous ground, I saw a Pipit dart out from beneath my feet, and alight again close by, in a manner which I was sure could only be that of a sitting hen. I had but to step off the grass-grown hillock on which I was standing, to see the nest ensconced in a little nook, half-covered by herbage. But the appearance of the eggs took me by surprise, they were unlike any I knew, of a brown colour indeed, but of a brown so warm, that I could only liken it to that of old mahogany wood, and compare them in my mind with those of the Lapland Bunting. However there was the bird running about so close to me, that with my glass I could see her almost as well as if she had been in my hand. That she was a Pipit was undeniable, and thoughts of a species till then unseen by me began to dawn upon my imagination. I replaced the eggs without disturbing the nest, and carefully marking the spot, we retired. In half an hour or so we returned, going softly to the place, and Mr. Simpson reaching his arm over the protecting hassock of grass, dexterously secured the bird in his hand as she was taking flight. I then at once knew, from her pale fawn-coloured throat, that the nest we had found belonged to a species which up to that time I believed had been known in Europe only as an accidental visitant,—the *Motacilla cervina* of Pallas, the *A. rufogularis* of Brehm.

A day or two later Mr. John Wolley returned from a Swan-upping expedition he had been making in the territories of our then imperial enemy. He told us that previous to his starting he had shot, somewhere in the neighbourhood of Wadsö, an example of a Pipit, which had puzzled him a good deal. The bird, which, during his absence, had been kept in a cellar, was produced, unskinned and still fresh, but unfortunately half eaten by mice. A very short inspection served to shew that it was a male of the same species as the hen we had, as above-mentioned, taken from the nest. Being too much injured to be preserved, it was reluctantly thrown away.

In a week's time we were quartered at Nyborg, a small settlement at the head of the Waranger Fjord. Here willows and birches grew with far greater luxuriance, even at the water's edge, than lower down the inlet. Some even attained to nearly twice the height of a man, and formed thickets, which, the intervening spaces being exceedingly boggy, were not easily explored. In this secluded spot we found our Red-throated friend not unplentiful. We could scarcely go out of the house without seeing one, and in the immediate neighbourhood we procured several more identified nests, making a total of five, and a fine series of nine birds, all, of course, in their breeding plumage. We had also abundant opportunities of watching their habits, and, above all, of contrasting them with those of the Titlark, (*A. pratensis,*) which was not uncommon in the district, and to which this species has been so unjustly annexed as a variety. The two birds had, according to our observation, an entirely different range; *A. pratensis* haunting a station less wooded (saving the expression) than that of *A. cervinus,* which latter we found at times feeding on the sea-shore—a habit we did not there notice the former to indulge in. No one with ears either could for a moment be in doubt about their respective notes. It is true that the full song of *A. cervinus* did not differ so strikingly from the more feeble performance of *A. pratensis* as does, for instance, the joyous burst of *A. arboreus*, but it had an unmistakable resemblance to the louder and perhaps harsher strain of *A. obscurus;* and in all cases was sufficiently characteristic for one to be quite certain as to the nature of the performer, even when the individual was not in sight. In a word, none of our party had any hesitation as to regarding *A. cervinus* as a *perfectly good species.*

I do not take upon myself a description of the specimens which I have had the pleasure of sending to Dr. Bree. A young bird, obtained at Mortensnæs, (between Wadsö and Nyborg,) July 16th., and as it was attended by its parents, (both of which were *well seen* by Mr. Wolley and myself,) could only have just left the nest, seems to differ only from the young of the Titlark in being of a ruddier complexion: a coloured drawing of it, made only a few hours after its death, is now before me. I have already mentioned what the eggs looked like, and it would be difficult in words to convey a better idea of them. All the nests I saw were simply built of dry bents, without any lining of feathers or hair.

I may however add that it was only in this restricted locality in East Finmark—between Wadsö and Nyborg —that we saw this bird, and I believe Mr. Wolley never met with it elsewhere, though a nest of unidentified eggs, brought to him, in 1854, from Nyimakka, ("v. p. 1066,") a settlement on the upper part of the Muonio river, *may* possibly belong to this species. At Stockholm I saw in the possession of Conservator Meves, the ingenious discoverer of the cause of the bleating noise made by the Common Snipe, a living Red-throated Pipit, which had been taken in a garden near that town, where, I believe, it not unfrequently occurs in its autumnal migration."

Middendorff expressly states that the bird he describes is the same as that of Keyserling and Blasius; and Mr. Newton's species is evidently that described by Middendorff, as *M. cervinus*. The male bird figured by the latter is less spotted on the breast, and the cheeks are more covered with russet.

Middendorff remarks, "This bird was found in both north and south Siberia. I shot a female in the Stonowoj Mountains, on the 26th. of May, consequently not on the passage. The rust-yellow of the Siberian specimen has a somewhat violet tint, (very similar to the colour on the breast of the Turtle Dove;) it covers the cheeks near the eyes, the throat, flank, neck, and upper part of the breast. It is only found in this plumage from May to July. These colours are on the upper part of the breast, sharply bordered with whitish yellow; on the belly, pencilled blackish spots. The back is very dark, without reddish or yellowish tints, but the narrow border of the feathers is whitish or greenish grey. The four first wing primaries are of almost equal length, and nearly as long as the longest tertial, as pointed out by Keyserling and Blasius. The inner half of the white outermost tail feather brownish, and there is a pointed three-cornered white patch on the tip of the inner web of the second feather; the rest of the tail feathers are black brown. When fresh the upper part of the beak was dark horn-colour; the under mandible at the point the same, but at its root light yellow; iris, dark chesnut brown; tarsus and toes brighter than iris, but the soles of the feet orange yellow."

The adult male sent to me by Mr. Newton, and marked "Nyborg, July 3-4, 1855, J. W. & N.," agrees in the main with the above description. It has however partly lost the rust-red on the cheeks, and the crop and flanks are more spotted. The feathers of the back, head, and nape are dark blackish brown, bordered with greenish grey, lighter on the head and nape; upper part of tail dark brown; underneath the throat and part of the cheeks and crop russet red, with a circlet of longitudinal dark spots on the crop. The abdomen, flanks, and under tail coverts creamy white, with a tinge of russet, the flanks being thickly covered with long longitudinal brown spots. Wings and tail underneath glossy mud-brown; tarsi and toes yellowish brown. The four first primaries of nearly equal length, the fourth shortest; tertials very long, the longest going beyond the end of primaries. Upper and lower wing coverts, as well as tertials, broadly bordered with light yellowish or creamy white. The tail is emarginate, dark brown above, and earthy brown below; first lateral quill white, with a dusky patch on lower half of inner web; second quill brown, with a triangular white patch at the tip, exactly like the drawing of these two feathers given by Middendorff.

The female is half an inch less in length, and has the upper plumage precisely the same as the male; below the russet is confined to the throat and cheeks, while the front of the neck and crop are thickly covered with rich brown longitudinal patches and spots, contrasting with the rich cream-coloured ground of the under parts; the under tail coverts have the light yellow russet or cream-colour more pronounced than the abdomen; upper mandible and anterior half of lower dark brown; basal half of lower mandible yellowish; tarsi yellowish; feet darker.

My figures of this bird are taken from specimens

kindly sent me by Mr. Alfred Newton; the female is the bird taken on the nest, described in Mr. Newton's notes. The egg is from a specimen sent me by the same gentleman.

Figured also by Middendorff, Sibirische reise, vol. 2; Gould, B. of E., pl. 140; Dubois, Ois. de la Belgique Liv., 51, pl. 97—a; Naumann, vol. 3, pl. 85, fig. 4.

INSECTIVORÆ.
Family MOTACILLIDÆ.
Genus ANTHUS. *(Bechstein.)*

WATER PIPIT.

Anthus spinoletta.

Alauda spinoletta,		LINNÆUS; S. N., 1766, vol. i., p. 288.
"	"	GMELIN, 1788. LATHAM; Ind., vol. ii., p. 495, 1790.
Anthus aquaticus,		BECHSTEIN; Nat. Deut., 1802, vol. i., p. 258.
"	"	TEMMINCK; Man., 1820 and 1840.
"	"	VIEILLOT; Dict., 1818, tome xxvi., p. 495, Faun. Fr., p. 180.
"	"	LESSON; Ornith, 1831.
"	"	BONAPARTE, 1838.
"	"	KEYSERLING AND BLASIUS, 1840.
"	"	SCHINZ; Faun., Europ., vol. i., p. 203.
"	"	SCHLEGEL; Revue, p. 35, 1844.
"	"	RICHARDSON AND SWAINSON; Faun. Bor. Amer., ii., 231, pl. 44.
"	*spinoletta,*	DEGLAND, 1849.
"	"	BADEKER AND BREHM, 1859.
Pipit spioncelle,		OF THE FRENCH.
Wasser pieper,		OF THE GERMANS.
Alouette pipi,		BUFFON.
Pispolada spioncella,		STOR.
Pipit spipolette,		VIEILLOT, Faun. Fr., p. 180.

Specific Characters.—Hind claw one tenth of an inch longer than the toe, and curved like the Tree Pipit; outer tail feather white below, with a dusky patch on the greater part of inner web; the superciliary ridge broadly white towards the occiput; top of the head and nape greyish olive; beak and feet black, the former strong. Length six inches and one fifth; carpus to tip three inches and a half; beak from gape three fifths of an inch, along ridge half an inch; hind toe three tenths of an inch; claw of hind toe two fifths of an inch; middle toe four fifths of an inch; claw of middle toe barely one fifth of an inch.

Owing to the term "*aquaticus*" having been applied to our Rock Lark, this bird has been confounded with it. It is, however, a very different species. It has also been confounded with *Anthus ludovicianus*, from which, however, it is said to be distinct by Zander, Brehm, and others. Mr. Morris, in his work upon British birds, has given the figure of "a Red-breasted Pipit," said to have been killed in Scotland, under the designation of *Anthus montanus*, Koch, which he states is synonymous with *Anthus spinoletta*, Bonaparte, *Anthus aquaticus*, Bechstein, *Anthus ludovicianus*, Lichenst, and *Alauda rufa*, of Wilson.

The bird which I now figure is not, however, the one given by Mr. Morris. It is the real *Anthus aquaticus* of Bechstein well and clearly described by Temminck in the last edition of the "Manual," and by Degland in his "Ornithology," in 1849. I should have preferred to have retained the term *aquaticus*, had it not been sometimes but erroneously applied to our Rock Lark, *A. obscurus*.

With regard to the distinction of the Water Pipit from the American "Red Lark," *Anthus ludovicianus*, we have the following diagnosis from Dr. Zander, (Cabanis' "Journal fur Orn.," 1853 and 1854,) and quoted by Baird in his "Birds of North America," to which my attention has been drawn by Mr. A. Newton. "*Anthus pennsylvanica, (A. ludovicianus.)* Specific characters.—Bill and feet blackish; longest tertial one line shorter than the longest primary. The light marking on the outer tail feathers shining white, and on the outermost one involving the half of the feather, —its shaft for the most part white. Body above olive green, the superciliary stripe yellowish."

Brehm (Bädeker's eggs,) describes the two birds separately, and he refers to the original description by Linnæus, as *Alauda spinoletta*, who pointed out as habitats the residence of this bird and not the American European straggler.

Assuming, then, the two birds to be distinct, and yet as closely-allied as the representatives of the two species of the Old and New Worlds so frequently are, the question arises, which of them is the bird which has been introduced into the British lists? Mr. Morris has given a figure of neither. His bird is evidently a specimen of *Anthus cervinus*. But Mr. Macgillivray distinctly describes with much clearness and at great length the American species, and he concludes by saying that the two birds shot near Edinburgh, are perfectly identical with the description taken by him from American specimens in Audubon's "Synopsis."

The "Water Pipit" or "Mountain Pipit," or as it was called by Latham, "Meadow Lark," is an inhabitant during the breeding-season and summer of the Swiss Alps, the Tyrol, the Pyrenees, and other high mountainous districts. In the autumn and winter it descends to the plains, and then gains its title to a "Water Pipit," by living along the course of rivers. It is found in Sweden, and in the mountains of Bavaria and Italy, and has occasionally, but rarely, been captured in Bulgaria. It does not occur in the north of Europe, keeping, according to Temminck, in the south and east.

"During their autumnal migration," according to Dubois, "they arrive in small flocks in the plains, moist meadows, and pastures, as well as on the banks of streams and ponds. They are very wild, and will only fly around any intruder while hatching their young. The male assiduously utters its rather simple song, during which it rises into the air, singing at first very slowly, and gradually increasing in intensity, 'tuigh, tuigh, tuigh,' and finishes when it comes down with a long 'si-si-si-si-si,' having its wings extended, and settling upon a shrub or stone always in an oblique line: it rarely sings when perched."

It feeds upon gnats, flies, and aquatic insects and their larvæ. From Bädeker's "European Eggs" I take the following:—"The Water Pipit builds its nest in moss, or under some overhanging stone. It is made of stalks, grass blades, and moss interwoven together, is well rounded in shape, and lined with grass and hairs. At the end of May it lays five or six eggs, which are tender-shelled, having a ground-colour harmonizing with the very soft and thickly-scattered spots and dots which cover the egg. Some appear brown, others grey, while others incline to a green. Many have a circlet of spots and hair-streaks at the larger end. One specimen has a bright grey ground, and is entirely covered with olive brown spots and dark grey dots; while in another the ground-colour is inclined to reddish, and has a few dark grey spots or dots on it.

The female watches clamorously, and sits constantly on the eggs. The young are hatched in fourteen days, and in their first plumage resemble the old birds in autumn. In autumn both young and old forsake their nesting-places, and betake themselves to brooks and rivers in mountain valleys. It is seldom found in Germany in winter."

The egg, from which my figure was taken, is one of two sent me by Mr. A. Newton, who received them from M. Nager Donazain. There is something incomplete in their history, but the character of the egg drawn corresponds with those in Bädeker's plate. The skin, from which my figure is taken, was also kindly sent me by the same gentleman.

In breeding-plumage the male has the top of the head and nape bluish grey, more or less mottled with the olive green tint of the back and rump. Wings and tail rich hair brown, the greater and lesser coverts of the former bordered with yellowish, which forms two bands across the wing. The superciliary ridge is white; cheeks grey; throat, sides of the neck, and crop, fawn-coloured, more or less mottled with bluish grey; the lower parts of the abdomen shaded off from this colour to dirty white; under tail coverts white. First four primaries of nearly equal length, but the third slightly the longest, and the first the shortest. The longest tertial does not reach the end of the longest primary by at least six lines. Beak, tarsi, and feet black; iris brown. In winter the fawn-colour of the lower parts is replaced by dirty white, with longitudinal spots more or less on the abdomen and flanks; the superciliary ridge is whiter, and the wing coverts are more broadly bordered with whitish grey. This applies to both male and female.

The female in breeding-plumage does not differ much from the male, except in having the fawn-colour more russet, and the superciliary ridge whiter.

The young of the year in their passage, according to Degland, resemble the adults in autumn, but the spots are more numerous and confluent below; the upper parts are of a darker brown; the tail is more emarginate, the two most external quills of each side are whiter, and the third has a small spot at its point; the beak and feet are of a clearer brown.

This bird is figured by Buffon, pl. enl. 661, f. 2, as *L'Alouette Pipi*. Naumann, pl. 85. Richardson and Swainson, in Faun. Bor. Amer., ii., 231, pl. 44. Roux, Ornith. Prov., pl. 192, in autumn plumage. Boutcille, Ornith. du Dauph., pl. 28, f. 5. Dubois, Oiseaux de la Belgique, part 65, pl. 95, in winter and breeding-plumage. The egg is also figured by Thienemann and Bädeker.

INSECTIVORÆ.
Family MOTACILLIDÆ.
Genus ANTHUS. *(Bechstein.)*

PENNSYLVANIAN PIPIT.

Anthus Ludovicianus.

Anthus ludovicianus,	BONAPARTE.
" "	WILSON; American Ornith., vol. ii, p. 184.
" "	AUDUBON; Synopsis, 94.
" "	BADEKER; Die Eier der Europ. Voegel. Description of plate 35, fig. 6.
Alauda rubra,	GMELIN. LATHAM.
" *ludoviciana,*	GMELIN. LATHAM.
" *rufa,*	WILSON.
" *pennsylvanica,*	BRISSON.
" *campestris,*	LATHAM.
Anthus spinoletta,	BONAPARTE; Sinop., p. 90.
" "	MACGILLIVRAY; Man. Brit. Birds, p. 169.
" "	AUDUBON; Orn. Biog., pl. 10.
Alouette aux joues brunes de Pennsylvanie,	BUFFON.
Pipi Farluzanne,	OF THE FRENCH.
Polar-Pieper,	BREHM.
Red Lark,	EDWARDS. PENNANT. LATHAM.
Brown Lark,	WILSON AND NUTTALL.
Field Lark & Louisana Lark,	LATHAM; Syn. ii, 376.

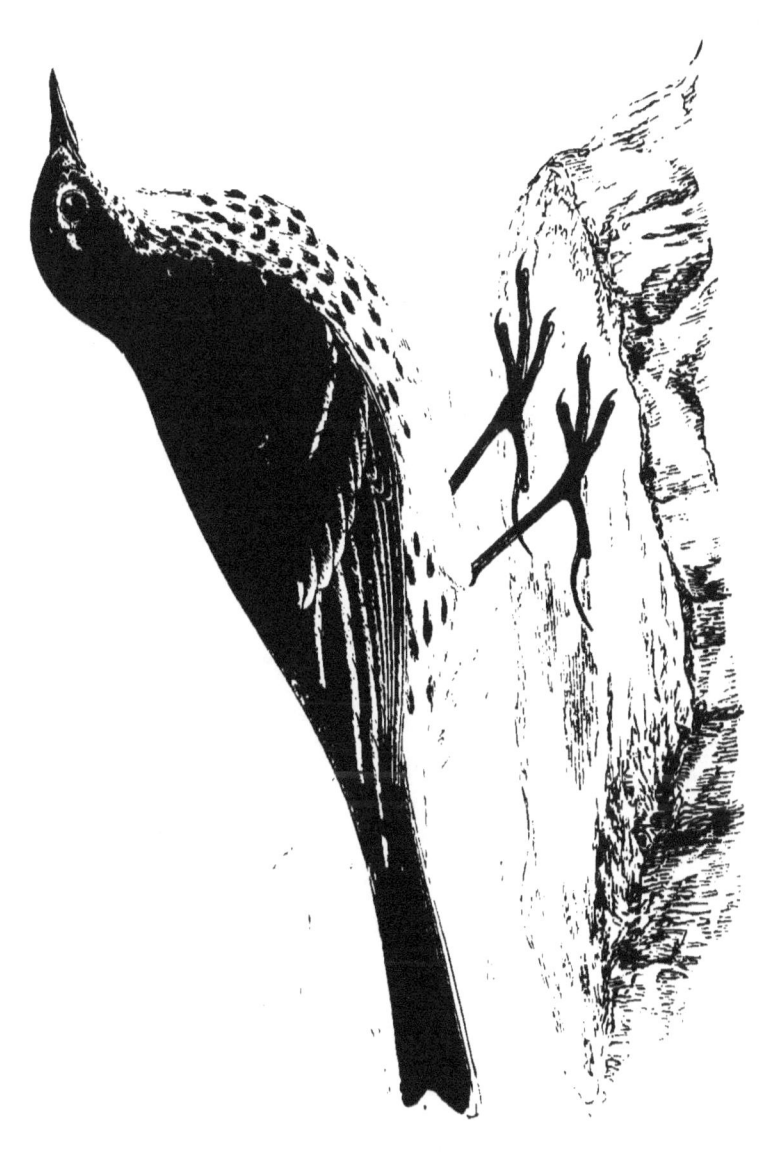

Specific Characters.—Superciliary ridge rufous yellow. Bill broad at the base, and the upper mandible has its dorsal line slightly declinated at the end. Hind claw one tenth of an inch longer than toe, and more curved than in *A. pratensis*, but less so than in *A. obscurus*. Upper plumage olive grey, and but slightly mottled; lower plumage tinged with rufous, and thickly spotted on the breast and flanks.

Length six inches; from carpal joint to tip, three inches and three tenths; tarsus nine tenths of an inch; hind toe seven twentieths of an inch; claw of hind toe four tenths of an inch. Beak along ridge half an inch, from gape three fifths of an inch. Tail two inches and seven tenths.

There has been so much confusion among the Pipits in consequence of the close affinities of several nearly allied species, and the usual misapplication of names which attend the designation of such families, that I have thought it best to give a figure and notice of this bird, although it has occurred in the British Isles.

I am the more induced to do this in the absence of any very correct figure of the bird, except that of Edwards, in the histories of our ornifauna; and also because it enables me to separate the synonymes, and to restore something like order in the diagnoses of the different species.

The Pennsylvanian Pipit, as I prefer calling it, in order to perpetuate the original description and notice of it by Brisson and Edwards, is, as its name expresses, an inhabitant of the New World, occurring accidentally in Heligoland and the British Islands. In the former it has been captured by Her Gätke, and is included in the list of birds published by him and Professor Blasius, in "Naumannia" for 1858. In the British Islands it has been taken frequently. Edwards says that in his time it was often taken about London, where it may probably still be found, mixed with *pratensis, arboreus*, or *obscurus*, among whom it passes unnoticed and unknown. More recently it was captured near Edinburgh by Professor Macgillivray, who has given a lengthened and very accurate description of the bird in his "Manual of Ornithology," as *Anthus spinoletta*. He records two specimens shot near Edinburgh on the 2nd. of June, 1824.

More recently still, Mr. Robert Gray, of Glasgow, has published an account of the occurrence of three specimens near that city. Mr. Morris has mentioned these captures in his work on "British Birds," vol. ii, p. 158, but the figure given was not taken from Mr. Gray's specimens which that gentleman informs me were unfortunately not preserved; but they corresponded in every particular with the description given by Mr. Macgillivray, clearly proving them to have belonged to the present species.

In America, the Pennsylvanian Pipit arrives in the Eastern and Central states from its breeding place in the north, about the middle of October.

In its habits, it is, as might have been inferred from its hind claw, intermediate between the Meadow and Rock Pipits of this country. It affects ploughed fields, running rapidly on the ground, but is also frequently found in the neighbourhood of rivers and marshes, and, as Nuttall has remarked, is especially fond of rocky coasts. "They utter a feeble note like 'tweet, tweet,' with the final tone often plaintively prolonged, and when in flocks they wheel about and fly pretty high, and to a considerable distance before they alight.

"It makes its nest in mountainous countries, even upon the sterile plains of those which are most elevated; more rarely in salt marshes, or in tufts of grass, on shelving rocks near the sea."—(Nuttall, Man. of Ornithology, vol. i. p. 452.)

Brehm remarks, in his notice of the egg in Bädeker's work, that it dwells in North America, in the inner half of the Polar circle, especially Greenland, travelling southward in the autumn, and sometimes wandering into Europe. It breeds in the high latitude above mentioned, forming a nest of moss, hair, and straw, lined with dry grass, stalks, or hair. It lays five or six eggs, which are very similar to those of the Meadow Pipit, though the spots are generally so indistinct, that they appear to be unicolorous grey, brown, or brownish red. They are sometimes found of a lighter colour, with markings displayed at wide intervals singly, and shining through a grey or bluish white ground. Some are marked with blackish brown hair streaks. The Polar Pipit has, like the Water and Rock Pipit, a double moult, but with the latter, it does not always put on the breeding plumage, retaining through the breeding season its autumnal dress.

The male and female have the upper parts of the body olive brown, the head, nape, and back, having a slightly mottled appearance; primaries brown; the secondaries and long tertials darker, and edged with light grey, forming two bands across the wing. Tail slightly emarginate, the two upper feathers olive brown, the others dark black brown, except the outer ones on each side, which are white, with half of the inner web dark brown; the second feather on each side slightly tipped with white on the outer web. The chin and superciliary ridge pale yellowish white, tinged with rufous; neck, crop, and flanks thickly covered with longitudinal brown spots, on a rusty white ground; abdomen and under tail coverts yellowish white, having a rusty tinge. Upper mandible, tarsi, and feet purplish brown; the lower mandibles yellowish brown, darker at the tip. The iris is said by Wilson to be a dark hazel.

When placed beside *A. pratensis*, there is a general resemblance between the two birds in size, colour, and markings. It differs, however, from *A. pratensis* in the following points:—The wing is half an inch longer, the hind claw shorter and more curved, the beak is stouter and broader at the base. Plumage of back is more uniform olive grey, and is less mottled, and on the lower part of the body there is a rufous tint more or less pronounced.

The above description is taken from a female specimen apparently in autumn plumage, and which is figured. It was kindly sent me by Mr. Sclater, the Secretary to the Zoological Society, to whom I have much pleasure in tendering my thanks for the assistance which he willingly affords me in the prosecution of this work. It is marked "Petaluma Car. Anthy. Baird e. mus P. L. Sclater, No. 5415."

It has been figured as *A. aquaticus* by Audubon, pl. 10, Orn. Biog., i, p. 49; Wilson, p. 89, pl. 42, fig. 4, (young.) The colours are much too bright in this figure —Edward's Gleanings, ii, 185, pl. 297.

I am sorry I have not an egg to figure. Brehm's very clear description renders it, however, unnecessary for me to copy any of the three figures he has given in plate 35, fig. 6. They are there represented as rather larger than those of *A. pratensis*, but thickly spotted with the same coloured dots on a similar ground.

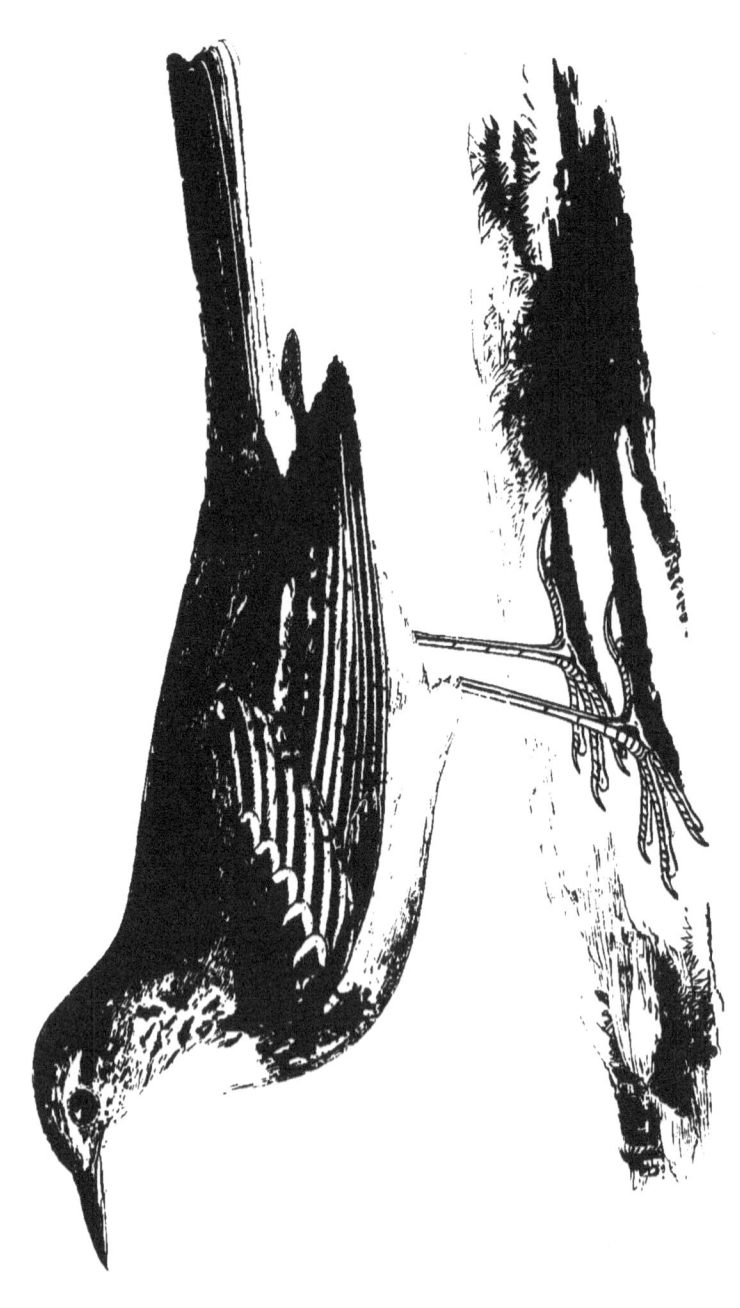

INSECTIVORÆ.
Family *MOTACILLIDÆ.*
Genus ANTHUS. *(Bechstein.)*

TAWNY PIPIT.

Anthus rufescens.

Alauda campestris,		BRISSON; Orn., v. iii, p. 349, 1760.
"	"	GMELIN; Syst., 1788.
Anthus campestris,		BECHSTEIN; Vog. Deut., 1807, vol. iii, p. 722.
"	"	MEYER AND WOLFF; Tasch der Deuts, 1810, vol. i, p. 257.
"	"	KEYSERLING AND BLASIUS.
"	"	SCHLEGEL. DEGLAND. DUBOIS.
"	*rufescens,*	TEMMINCK. SCHINZ.
"	*rufus,*	VIEILLOT.
Pipit rousseline,		OF THE FRENCH.
Brachpieper,		OF THE GERMANS.

Specific Characters.—The claw of the hind toe shorter than the toe, and but slightly curved. The two most lateral tail quills white, the inner barbs smoky brown for three quarters of their length. Slight appearance of a moustache. Length six inches six lines; carpus to tip three inches; tail three inches.

THE Tawny Pipit inhabits the southern and temperate parts of Europe, although, as we are informed by Dubois, it is sometimes found in Sweden and Finland. It is found during its passage also in the north of France. It is a common bird in Lorraine, Sicily, and Provence, especially in the department of the Var and the Basses Alpes, where it stays from April to September. It is found in the temperate parts of Asia, Nubia, and Egypt, and is included in Mr. Tristram's list of birds, observed by him in Palestine, and on the *Hauts Plateaux* of Northern Africa.

The following account of the habits of the Tawny Pipit is from M. Dubois' "Birds of Belgium," a work which has the merit of giving in the short notice of each bird, a considerable amount of information of its mode of life:—

"The 'Pipit des Champs' lives by preference on extensive dry plains, where hardly a plant or a tree is to be found; it loves much to live in large flocks, shunning the high grass and bushes. It is almost always on the ground, sometimes perched upon a hillock, a stone, or a bush; rarely it is found on trees. It is very shy, lively, and coy in its movements. The male has a singular song, composed of short, uniform, and melancholy tones, which it utters while flying; in autumn it congregates in small flocks in fallow fields.

It feeds upon small coleoptera, spiders, and many other insects and their larvæ. Its nest is found on the ground, in slight hollows sheltered by a bush; it is composed of blades of grass and moss, the interior being lined with hair and rootlets, and it contains four or five eggs. The young quit the nest before they can fly, for they can always run sufficiently well to hide themselves in the grass, corn, or brushwood."

According to Degland, "it builds in sand-beds, in fields on the ground, under the cover of a stone, or a clod, or a small bush; sometimes on the mountains, in the

crevices of rocks. It lays four to six eggs, which are dirty white, greyish, or greenish, covered with small spots more or less abundant, greyish russet brown, or russet green, and sometimes finely spotted with greenish or brown red. It prefers to live in uncultivated and stony or dry hilly places, covered with heath and thyme. It runs quickly, and with grace, and very rarely perches upon large trees. Its cry is very like that of the Short-toed Lark. It lives principally on neuropterous insects."

The male in breeding plumage has all the upper parts of a light russet brown; wings and two upper tail feathers darker; the edge of all the upper feathers more or less bordered with light grey; six middle tail feathers deep brown, the two lateral quills on each side white, with a longitudinal patch of black brown on the inner webs; a few dusky feathers form a small moustache beneath the eyes and auditory orifices. All the lower parts of the body bluish white, lighter at the lower part of abdomen and under tail coverts; beak blackish brown above, yellowish below; tarsi and feet yellow; iris brown.

The females are like the males in all seasons, but the colour is lighter, and they are somewhat mottled about the crop.

The young before the first moult are, according to Degland, of a browner plumage above, with the feathers bordered with clear russet; crop and flanks marked with a greater number of more elongated spots.

My figure is a male in the breeding plumage, shot by Mr. Tristram, at Kif Laki, April 22nd., 1857. The egg from which my figure is taken was also kindly sent me by the same gentleman, who remarks, "This egg is very variable, though not so much so as that of *B. arboreus.* Some of my specimens approach those of the Pied Wagtail; in others the russet spots are as large, thick, and bright, as in *Sylvia galactodes*, which egg this variety exactly resembles."

Also figured by Buffon, pl. El. 661, f. 1, as La rousseline: Pennant, British Zoology, p. 95, pl. Q, fig. 4, as "Willow Lark;" Frisch. Brachlerche, pl. 15, 2 A; Naumann, pl. 84, fig. 1, Nat. Neue. Ausg.; Vieillot, Faun. Franc, pl. 78, fig. 2 and 3; Roux, Ornith. Prov., pl. 191, f. 1, adult, fig. 2, head of young; Bouteille, Ornith. du Dauph., pl. 29, f. 1; Gould, B. of E., pl. 139; Dubois, Ois de la Belgique; Livr, 61, pl. 96 A.

ORDER IV.—GRANIVORÆ.

Family ALAUDIDÆ. *(Bonaparte.)*

Genus ALAUDA. *(Linnæus.)*

Generic Characters.—Beak sub-conic, mandibles of equal length, the upper one convex, and more or less curved or straight. Nostrils basal ovoid, covered by small feathers directly forwards. Feet.—Three toes in front, one behind, the middle joined at its base, with exterior claw of hinder toe straight, or nearly so, and generally longer than the toe. Wings.—First quill obsolete, or nearly so, the second shorter than the third, which is the longest; two of the secondaries as long as the primaries. Feathers of the head more or less elongated, and capable of erection.

SECTION I.—LARKS WITH ARCHED BEAKS.

Genus CERTHILAUDA. *(Swainson.)*

Beak as long or longer than head, slightly arched.

BIFASCIATED LARK.

Alauda bifasciata.

Alauda bifasciata,	LICHSTENSTEIN; Cat. 1823, p. 27.
" "	TEMMINCK; Man., 1835.
" "	SCHINZ. SCHLEGEL.
" "	DEGLAND.

Certhilauda bifasciata,	BONAPARTE; 1838.
Alæmon desertorum,	STANLEY.
" "	KEYSERLING & BLASIUS; 1840.
Alouette double-bande,	OF THE FRENCH.
Zweibindige Lerche,	OF THE GERMANS.

Specific Characters.—The false primary about one-third the length of the first, which is shorter than the fourth. The two longest of the greater wing coverts very nearly as long as the fifth primary. Tail long, with the outer web of the most lateral quills white; posterior claw as long as the toe. A broad black band across the white secondaries. Length eight inches six lines. —TEMMINCK, which is exactly the length of the female specimen sent me by Mr. Tristram, and which is figured.

THE English naturalist who confines his attention to his own fauna, a habit, the breach of which will afford him great pleasure and instruction, will be struck with the difference between the Bifasciated Lark and one or two others which I shall have to bring before his notice, and the well-known graceful forms of our Skylark or Woodlark. But the family is well linked together by similarity of structure and habit, which we shall see as we proceed.

The Bifasciated Lark is an inhabitant of Andalusia and Candia, and has been seen occasionally in Sicily and the south of France. Dr. Leith Adams informs me that this bird is also found in the deserts of Western Asia and Scinde. Its real home, however, is in the north of Africa, where its habits have been observed by Mr. Tristram, and by whom the bird which I have figured was shot. I extract the following from one of Mr. Tristram's papers on the ornithology of Northern Africa, in that excellent and useful work, the "Ibis," vol. i, p. 426:—"The Bifasciated Lark is universally distributed throughout the whole of the true desert. Unlike its congeners it seems to be a most solitary bird, and seldom, except in the breeding-season, have I seen two together. But a day rarely occurred when we did not obtain a few specimens on the march; and indeed this game formed our principal and favourite animal food. Although its uniform of inconspicuous drab renders it most difficult of detection on the ground, its restless habits soon attract attention. The moment it extends its wings the broad black bar across the snow-white secondaries attracts the eye, and renders it an easy mark. At first sight it reminded me much of a Plover, in the manner in which it rose and scudded away. Indeed there is nothing of the Lark in its flight, except in early morning, when I have watched it rise perpendicularly to some elevation, and then drop suddenly, repeating these gambols uninterruptedly, over exactly the same spot for nearly an hour, accompanying itself by a loud whistling song. It runs with great rapidity, and it requires no little spread of foot to capture a broken-winged victim. In the stomach of those I opened I found small coleoptera, sand-flies, and hard seeds.

There is something very graceful in all its movements, and the distinct markings of its wings, and the expansion of its long black tail, render it really a beautiful bird when flying.

The egg is very large—twelve lines by eight; the ground colour like that of *C. duponti*, but the brown blotches smaller, and far more closely distributed, especially towards the broader end. It would not be easy to select it out of a series of some varieties of *Lanius excubitor*."

Mr. Tristram has described in the same paper another

Certhilauda closely allied to this, under the name *C. salvini*. The sterna drawings of which are given, are certainly very different, even in important osteological characters. This bird is shorter by one-fifth of an inch, more slender, and has a broader white band on the secondaries than *C. bifasciata*. Mr. Tristram suggests, however, it may be only a local race, although this idea is rather negatived by the fact that both he and Captain Loche had independently arrived at the conclusion, that these were two species, and that the smaller one, though confined to the southern and south-eastern districts, never being found in the central or western, yet did not supplant the common bird in the districts where it occurred.

These observations lead Mr. Tristram into a very interesting discussion of the now exciting question of the variation in species. Though tempting, I have not room in this work to follow him in his remarks, but I must refer the reader to the first volume of the "Ibis," p. 429, et seq.

I may remark, however, that while Mr. Tristram thinks that observations he made on the Larks and Chats of North Africa, illustrate the views of Mr. Darwin and Mr. Wallace upon this subject, he distinctly repudiates the possibility of such a law acting beyond the sphere of species and race. "I do not," he says, "for a moment mean to imply that such birds as *Rhamphoceris clot-bey* have been developed out of any known European form, or that we are to presume so far to limit Creative Power, as to endeavour to explain the growth of desert species universally by the development of individual peculiarities."

It will be well for science and themselves, if all naturalists will stop at the boundary line thus drawn by Mr. Tristram. That species do vary, no naturalist denies. That they do this beyond the peculiarities by which the species is recognised, *no one has ever yet proved.* "Naturam expelles furca, tamen usque recurret."

The adult male and female of the Bifasciated Lark do not differ in their plumage. The upper parts are of a light chesnut, or isabelle colour, tending more to grey on the top of the head and nape, and the upper tail feathers being darker chesnut, with lighter borders. The auriculars are mixed chesnut and black, and there is a slight white superciliary ridge. Throat white; neck, abdomen, and under tail coverts, light creamy chesnut, with a row of dark spots where the white of the throat joins the crop. The primaries dark brown, having a white band commencing slightly on the second. The secondaries are white, with the dark brown of the primaries extended across in the form of a band, occupying their middle third. Tail same colour as primaries, except the two upper and two or three lateral ones, which are as stated in the specific diagnosis. Beak and feet yellowish; iris brown.

The young, according to Degland, have the head and neck greyish, with each plume marked with brown the length of the shaft; auricular region almost entirely white; crop more marked with black spots, and the colours of the plumage more strongly marked above and below.

My figures of the bird and its egg are from specimens kindly sent me by Mr. Tristram. The bird, which is a female, was shot at Wednça, December 10th., 1856. It is also figured by Temminck et Laug. pl. col. 393. Gould, pl. 168; Cretschm, voy. de Rüppell, pl. 5.

GRANIVORÆ.
Family ALAUDIDÆ.
Genus ALAUDA. *(Linnæus.)*

DUPONT'S LARK.

Alauda dupontii.

Alauda dupontii,	VIEILLOT; Faun. Fr., p. 173; 1828.
" "	TEMMINCK; Man., 1835.
" "	LESSON. SCHINZ.
" "	SCHLEGEL. DEGLAND.
" ferruginea,	MUHLE (?) Orn, Griech, p. 35.
Certhilauda duponti,	BONAPARTE; Consp. Av, p. 246.
" "	TRISTRAM; Ibis, vol. i, p. 427.
" "	LOCHE; Cat., p. 85.
Alouette Dupont,	OF THE FRENCH.
Dupont's Lerche,	OF THE GERMANS.

Specific Characters.—Beak as long as the head, and distinctly curved. Two outer tail feathers white, with a dusky band on the inner web; the second brown black, with the outer web white. Hinder claw shorter than the toe, and distinctly curved.

Length of male sent me by Mr. Tristram, which is figured, six inches and three quarters; from carpus to tip four inches; tail two inches and a half; beak seven eighths of an inch; tarsus one inch; hinder claw three eighths of an inch; hinder toe half an inch.

This bird has been considered by many ornithologists as a variety of the Skylark, and Keyserling and Blasius have even described it as a monstrosity. Homeyer, in Cabanis, Journal, Heft. 3, 1859, p. 204, speaks of it doubtfully as a species, and gives in the same article some excellent advice about the too prevalent habit of species making. In "Naumannia," part 3, 1858, Professor Blasius, however, after giving the various opinions which have been held about this species by authors, states that he has at length received a specimen from Algeria, which he admits is that described by Temminck, and must be considered as distinct.

With the birds before me, I cannot help expressing surprise that *A. dupontii* should ever have been confounded either with the bird last described, or with the Skylark. It is perfectly distinct from each, as the specific characters above will shew, and I have much pleasure in being able to add something to its natural history, not only by giving a figure of the bird shot by Mr. Tristram, but also a drawing of the egg, which I believe has never before been figured, and which rare and precious specimen that gentleman was kind enough to trust to me for illustration.

Dupont's Lark is found in Syria, some parts of Barbary, and in the south of Spain. Its occurrence in the centre of Europe is, however, only accidental. Degland says it is frequently found exposed for sale in the markets of Marseilles. The real home of this interesting species is, however, among the sandy deserts of Northern Africa, where its habits have been observed by the Rev. H. B. Tristram and Captain Loche. From the description of the former in the "Ibis," vol. i., p. 427, I quote the following:—

"This elegant and delicately-marked bird, a link between Galerida and Certhilauda, beautifully illustrative of the gentler gradations by which Nature glides from one type to another, is, I believe, the very rarest of all the Larks of the Sahara. I found it only in the far south, in the Wed Nça, at which place it was also obtained by Captain Loche a few months afterwards. Neither of us ever saw more than two or three pairs. The white outer tail feathers give it the appearance at first sight of our common Skylark, for which indeed it passed with my companion, who was the first to shoot it. Captain Loche obtained a nest of four eggs, one of which he kindly presented to me. As might have been expected, the eggs differ much from the typical characteristics of the Lark. They are very round, nine lines and a half by eight, of a soiled white colour, with pale brown blotches sparingly scattered over the surface, bearing a strong resemblance to small varieties of *Lanius excubitor*, but with an ivory-polished surface."

It is quite clear that all the doubts raised as to the specific distinctness of this bird, have arisen from an imperfect acquaintance either with its skins or habits.

The adult male in winter plumage (Mr. Tristram's specimen is marked December, 1856,) has all the upper parts of the body a rich brown, of lighter and dark shades, variegated with greyish on the edges of the feathers; top of the head darker, with a greyish longitudinal band across the vertex, and a similar one mottled grey and black, forming a kind of collar at the nape and round the neck; ear coverts clear brown, with a light grey patch above the eyes, and laterally on each side of the base of the upper mandibles; primaries and secondaries dark brown, with light chesnut edges; two upper tail feathers and upper tail coverts light chesnut brown, darkest in the centre; the first lateral tail

feather white, with black brown internal edge; the second black brown, with a white external edge; the six central feathers dark blackish brown. The under parts are of a dirty white, thickly spotted on the throat with dark brown longitudinal marks, and on the cross and flanks with the same shaped spots of russet brown; feet, beak, and iris brown.

Temminck says that the young differ from the adult by the large borders of clear isabelle colour, which mark all the feathers of the upper parts of the body; the black spots of the inferior parts are larger than in the adult. It is only seven inches long.

My figure of this bird is from a specimen sent me by Mr. Tristram, marked "Waregla, Dec., 1856, ♂." The egg is the one alluded to in the quotation I have made from that gentleman's paper in the "Ibis."

The bird has also been figured by Vieillot, Faun. Franc., p. 173, pl. 76, fig. 2; Roux, Ornith. Prov., vol. i, p. 285, pl. 186; Werner, Atlas du Manuel.

GRANIVORÆ.
Family *ALAUDIDÆ.*
Genus ALAUDA. *(Linnæus.)*

SECTION II.—LARKS PROPERLY SO CALLED.
Beak rather slender, and nearly straight and conical.

DESERT LARK.

Alauda isabellina.

Alauda lusitania,	GMELIN; Syst., vol. i, p. 798.
" "	LATHAM; Ind., vol. ii, p. 500.
" "	DEGLAND.
" *isabellina,*	TEMMINCK; Man., 4th. part, 1840, p. 637.
" "	SCHLEGEL, 1844.
Annomanes isabellina,	BONAPARTE; Consp. Avium, p. 244.
" "	TRISTRAM; Ibis, vol. i, p. 422.
" "	LOCHE; Cat., p. 83, No. 159.
Alouette isabelle,	OF THE FRENCH.
Isabellfarbige Lerche,	OF THE GERMANS.

Specific Characters.—False primary, one third the length of the first true one, which is shorter than the next four, and about equal in length with the sixth. Plumage beautiful isabelle or rich almond-colour. Hind claw slightly arched, and about the same length as the toe. Length of male specimen sent me by Mr. Tristram, six inches and a half; carpus to tip four inches; tail three inches; tarsus four fifths of an inch; toe three tenths of an inch; claw three tenths of an inch; beak seven tenths of an inch.

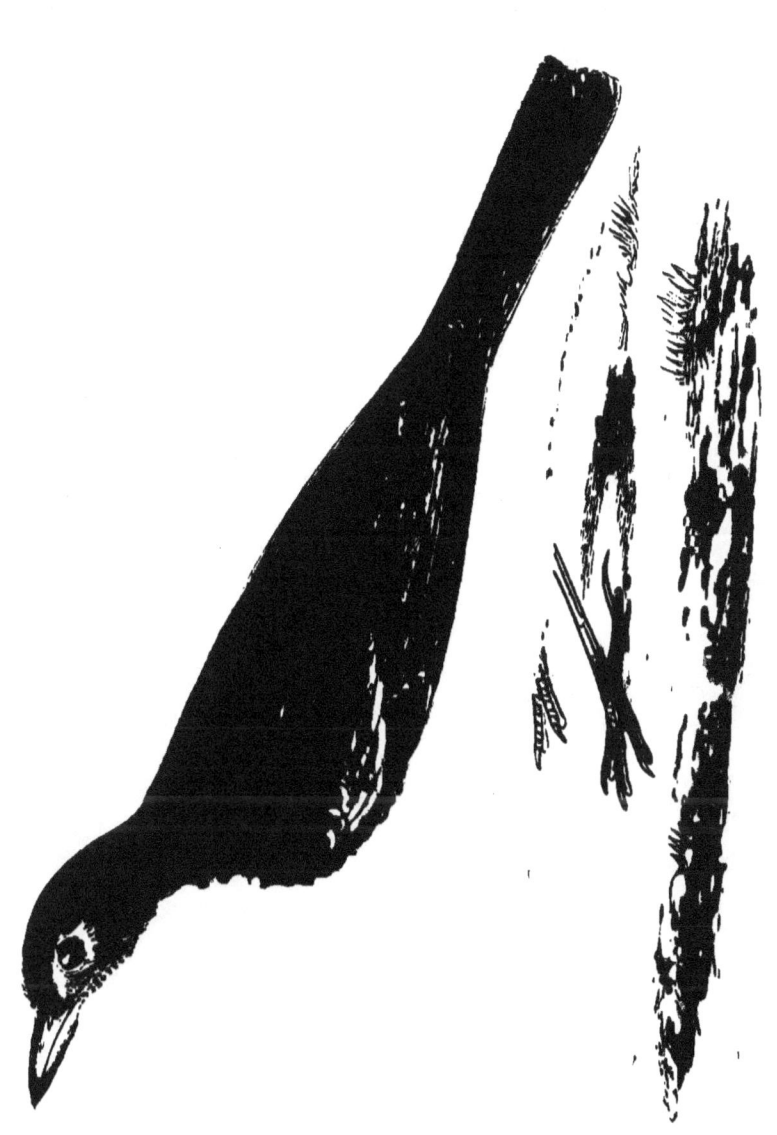

THERE is likely to be some confusion among the Desert Larks, in consequence of the adoption of similar names to designate different birds. Thus the present bird is called the Desert Lark, but Mr. Tristram gives the names of "Pale Desert Lark" and "Little Desert Lark" to two other North African species, while we have *C. desertorum* applied to the Bifasciated Lark. Then again, while Schlegel, Degland, and others refer the present bird to the *Alauda deserti* of Lichtenstein, Captain Loche, in his "Catalogue of Algerian Birds," following Bonaparte, makes the latter a distinct species, under the name of *Annomanes deserti*. Then again we have the name *Annomanes isabellina* applied to the subject of the present notice by Prince Ch. Bonaparte, which he gives to a closely-allied species the name of *Galerida isabellina*. Temminck describes our bird as *Alauda isabellina*, while Rüppell gave the same designation to the *Galerida isabellina* of Bonaparte. It must therefore be strictly borne in mind that the species found in Europe is the *Alauda isabellina* of Temminck, and the Desert Lark of Tristram.

This beautiful and elegant species was first described as European by Temminck, in the last edition of his "Manual," in 1840. Its European localities are Greece, south of Spain, and Portugal. It inhabits also Egypt, Arabia, and the north of Africa.

For a knowledge of its habits, hitherto recorded as unknown, we are indebted to Mr. Tristram, ("Ibis," vol. i, p. 422,) who writes:—*A. isabellina*, Temminck, occurs first on leaving the Hauts Plateaux in small numbers, but is more plentiful further south, inhabiting the open plains, where it is difficult to conceive how it finds subsistence. Its lateral range is wide. I have obtained it from the frontiers of Morocco to

Arabia Petræa. It is sedentary, and breeds both in the Algerian Sahara and in the wilderness of Judæa, in both which localities I have taken the nest, neatly formed of grass, in a depression under a tuft of weeds, and with four eggs, in size nearly equal to those of *Alauda cristata*, but never so elongated; measuring eleven lines by eight, of a rich cream-colour, blotched especially towards the large end with brown and red spots. In its habits this very distinct species exhibits, so far as I am aware, no distinctive peculiarities, living in small flocks, and poising itself in the air like its congeners. Its notes are few, though not unmelodious; but its song will bear no comparison either in volume or sweetness with that of the Skylark. It varies considerably in size, but its average length is about six inches and a half."

Dr. Leith Adams considers this bird as probably identical with *M. phœnicuroides*, Blythe, "I. A. S. Beng.," xxii, p. 583. It is found in Scinde and Cashmere. Dr. Adams gives the following measurements of the Indian species:—"Length about six inches; wing three inches and one-sixth; first primary one inch and one-eighth, being an inch and five-eighths shorter than the second; the second is a quarter of an inch less than the next three, which are equal; tail two inches and three-quarters. Bill to gape five-eighths of an inch; tarsus seven-eighths of an inch; hind claw five-sixteenths of an inch. Legs brown."

The male and female have the upper parts of a beautiful glossy dark fawn-colour, very much like that of our Almond Tumblers. The feathers shine and decompose the light like shot silk. Primaries and tail feathers brown, but bordered more or less deeply with the prevailing isabelle tint; below, the colour, though the same, is lighter, and the throat is whitish, mottled with dusky spots. Under wing coverts and part of the inner webs of wing feathers below, rich silky dark fawn; ends of primaries brown. There is the usual tendency to form a crest of the head feathers. Beak yellowish horn-colour; feet, legs, and iris, clear brown.

The young before the first moult, according to Degland, have the colours brighter, with the feathers of the upper parts of the wings and tail bordered with grey.

My figure of this bird is from a female specimen shot by Mr. Tristram, in the Northern Sahara, December 2nd., 1856. The egg is also from a specimen taken by the same gentleman in that locality.

The bird has been figured by Temminck and Laugier, pl. color. 244, fig. 2, from an Arabian specimen.

GRANIVORÆ.
Family ALAUDIDÆ.
Genus ALAUDA. *(Linnæus.)*

SECTION III.—LARKS WITH THICK BEAKS.

BLACK LARK.

Alauda tartarica.

Alauda tartarica,		PALLAS; Voy, 1776.
"	"	GMELIN; Syst., 1788.
"	"	LATHAM; Ind., 1790.
"	"	TEMMINCK; Man., 1820.
"	"	VIEILLOT. CUVIER. LESSON.
"	"	SCHINZ. SCHLEGEL.
"	"	DEGLAND.
"	*mutabilis,*	GMELIN; Syst., 1788.
"	*nigra,*	FALK; p. 796.
Melanocorypha tartarica,		CH. BONAPARTE; List, 1838.
"	"	KEYSERLING AND BLASIUS, 1840.
Calandra nigra,		DUBOIS; Ois. de la Belg. Liv. 73, p. 102 a.
Alouette nègre,		OF THE FRENCH.
Steppenlercher,		OF THE GERMANS.

Specific Characters.—No false primary; the three first true ones nearly equal; the fourth eight tenths, and the fifth seven tenths, of an inch shorter than the third; claw one fifth longer than toe.

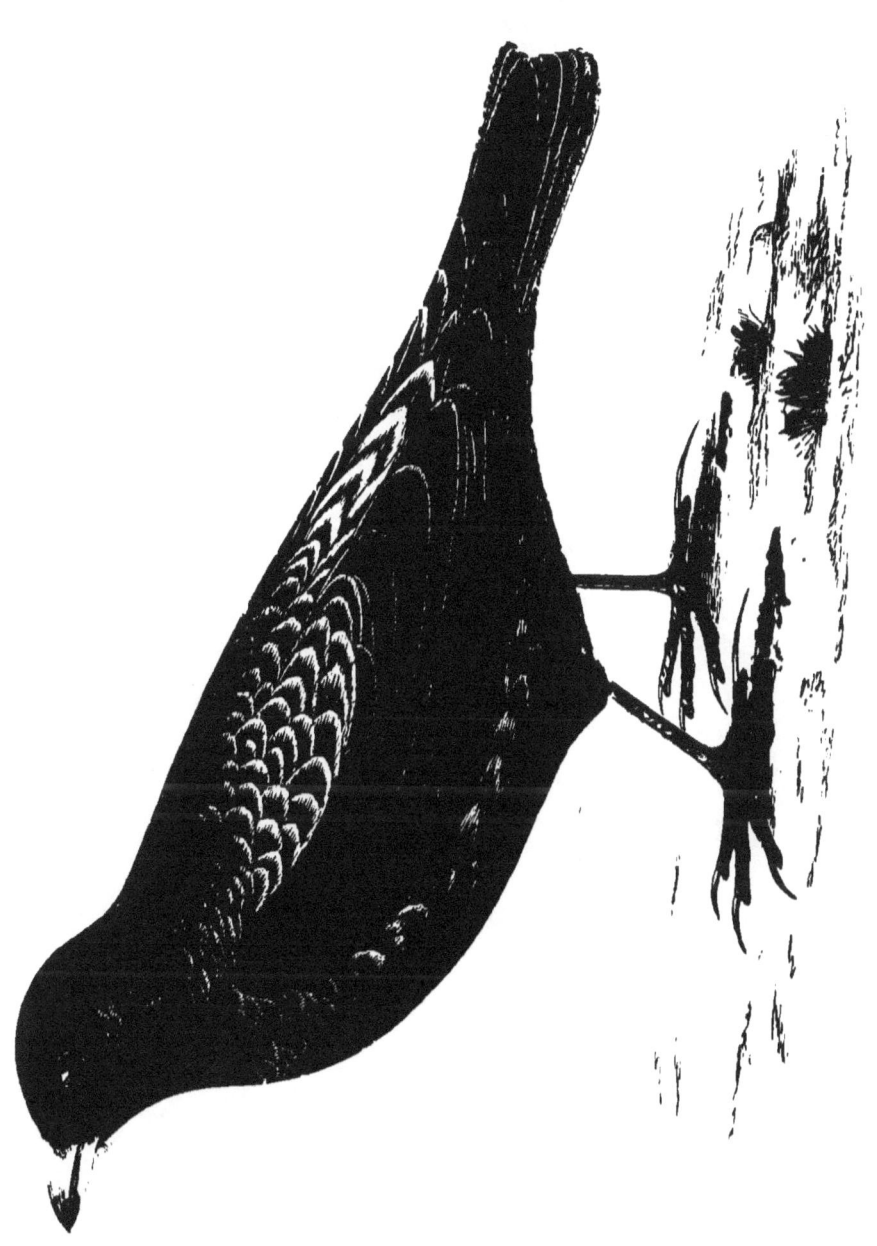

Beak one tenth of an inch longer than broad. Plumage black in spring, yellow grey in the autumn, with the wings and tail black.

Length of male specimen sent me by Mr. Tristram, seven inches and three quarters; carpus to tip five inches and a quarter.

The Black Lark is an inhabitant of northern climes. In Europe it is found in the precincts of the Wolga and Istych Rivers in Russia. It has been captured rarely and accidentally in Germany, and still more rarely in Belgium. On the authority of Dubois we have the record of one being trapped in the neighbourhood of Brussels, in 1850, and which he found in the market for sale. The person who caught it assured M. Dubois that there was a large flock, but he was only able to get one. M. Croegaert is also quoted by M. Dubois as having taken one in the neighbourhood of Anvers, in 1852, and kept it alive during several weeks. According to Pallas it is found in the wild and barren wastes of Tartary, between the Volga and Iaïk Rivers, whence it emigrates during the winter. It occurs also in the Steppes of Asia, and in the south of Africa.

"This bird," says M. Dubois, "inhabits during the summer in large flocks, the most extensive and infertile plains, where they may be seen from time to time on the sand-hills. They emigrate in autumn, and arrive during the rigorous winter at the villages and towns on their route, whence they penetrate to the interior. They return to the Steppes during the earliest days of spring. Their flight is to short distances, not very quick, and rather low. It is in general a careless bird; its song is not very good, and it generally sings seated upon a hillock; its call-note is heard (very rarely) when it rises into the air. Its nourishment is insects and their larvæ, worms, and seeds; it makes its nest upon the ground in a little excavation, and knows very well how to hide it, in spite of the great aridity of the soil. The nest is made without art or skill; it is composed of blades of grass, roots, and moss: in the interior small rootlets, and sometimes feathers. It contains from four to five eggs."

The bird differs much in plumage at different seasons and ages. In summer the entire plumage of the male is black; beak yellow, with the point brown, and in the breeding dress in spring it is, as represented in my figure, black, with the feathers of the back, rump, and flanks more or less bordered with white. In the autumn it is yellow grey, with scale-like spots, (according to Degland,) on the crop; stomach, wings, and tail black, the quills of both wing and tail bordered with grey white.

The female has, according to Temminck, all the plumage of a paler black, with the forehead greyish, and all the feathers of the neck, of the throat, and crop finely bordered with grey.

The young resemble the female, but the plumage has more of a brown shade, the brown of the feathers broader and more yellowish, the tail and wing quills being bordered with the same colour.

My figure is taken from a male specimen, from the Volga, sent me by Mr. Tristram. The egg is from Thienemann.

The bird is also figured by Buffon, pl. enl. 650, f. i.; Gould, B. of E., pl. 161; Vieillot, Galerie des Oiseaux, vol. i, p. 259, pl. 160, adult male; Gmelin, Nov. Comm. Petrop. xv., p. 479, pl. 23, fig. 2; Werner, Atlas du Manuel, pl. lith. of the young of the year.

GRANIVORÆ.
Family ALAUDIDÆ.
Genus ALAUDA.. *(Linnæus.)*

CALANDRA LARK.

Alauda calandra.

Alauda calandra,		BRISSON; Ornith., vol. iii, p. 352. 1760.
"	"	LINNÆUS; S. N. 12th. edit. 1766, vol. i, p. 226.
"	"	GMELIN; Syst., 1788.
"	"	LATHAM; Ind., vol. ii, p. 496. 1790.
"	"	MEYER AND WOLFF; Tasch der Deuts. p. 261, 1810.
"	"	VIEILLOT; Dict et Faun. Franc., 1816.
"	"	TEMMINCK; Man., 1820.
"	"	LESSON; 1831.
"	"	SCHINZ; Eur. Faun., 1820.
"	"	SCHLEGEL; Revue, 1844.
"	"	DEGLAND; 1849.
Melanocorypha calandra,		CH. BONAPARTE; List, 1838.
"	"	KEYSERLING & BLASIUS; 1840.
"	"	MÜHLE; 1844.
"	"	SALVIN & TRISTRAM; Ibis, 1859.
"	"	LOCHE; 1858.
Calandra bimaculata,		DUBOIS.
Alouette calandre,		OF THE FRENCH.
Kalander Lerche,		OF THE GERMANS.

Specific Characters.—Beak compressed, and the upper mandible curved and overlapping the under one, so as to form a sharp point—as long again as broad. First primary the longest, the second nearly as long, and about the same difference in size between that and the third; fourth an inch shorter than the third, and about the same difference in size between the fourth and fifth and the fifth and sixth. All the tail feathers except the two upper ones tipped with white, and the two laterals almost entirely of that colour.

Length eight inches; carpus to tip five inches and three tenths; beak three-quarters of an inch; hind claw three fifths of an inch; hind toe two fifths of an inch.

The Calandra Lark, one of the most conspicuous species of the genus, is also, perhaps, the best known of this section, being very common in many parts of the south of Europe. It is found in Italy, Portugal, and Spain, Roman States and Sicily, Sardinia, south of France, Greece, the Crimea, and the Steppes generally of the south of Russia. It is observed rarely in Germany, and, according to Temminck, never in Holland, but it is included, figured, and described by Dubois among the birds of Belgium.

In the north of Africa it is mentioned by Mr. Tristram as swarming on the coasts, but scarcer in the interior, so as hardly to lay claim to the Sahara as a locality. It is also included by that gentleman among the birds of Palestine, (Ibis, vol. i.) It is plentiful in Turkey, and in the Steppes of Southern Asia.

Like most of its tribe, except our Skylark, the Calandra seems to prefer wild and desolate plains to cultivated ground for its residence. In other respects, however, its habits are very similar. Dubois remarks, "They often fly together in flocks, and have a clear, beautiful, and varied song, which is heard as often when flying as in repose; notwithstanding the perfection of their song, it is impossible for amateurs to keep them in their homes, their voice is too loud. When taken young they may be taught to imitate the voices of all kinds of birds. They often mix together the notes of Thrushes, Finches, Tits, Quails, Linnets, etc., and will even imitate the croaking of the frog."

Captain Blakiston, in his description of the Birds found by him in the Crimea, ("Zoologist," 1857, p. 5509,) gives an amusing account of his first meeting with this bird, which will bear quoting:—

"I ask any naturalist, is there any pleasure in observing a new species for the first time? Surely you have a peculiar feeling within you; you eagerly wish for a specimen; and I will answer for it that you do not rest till you have obtained one. Suppose that you are a field ornithologist, you take the first opportunity, and although the weather is cold and windy, with snow covering the ground, you trudge off with your fowling-piece to where you observed the birds. I did this on the 2nd. of January, and found the bird I was in search of on the Karani hill, within sight of Sebastopol. I soon procured a couple, and after waiting in the snow behind an old bit of a wall for some time, knocked over six more at a shot; they were Larks, but the largest I had ever seen. I had studied Yarrell well when making out the Short-toed Lark. What could they be? To make sure, however, I turned over the leaves again that evening in my hut, but they were not there. I was at a loss. 'However,' I said to myself, 'they are Larks, but somewhat approaching to Buntings.' So I marked them in my journal as specimens of 'large Lark,' and noted the measurements and other points. This was the Calandra Lark, *(Alauda calandra,)* as I afterwards learned from England."

The Calandra Lark nests on the ground among lucerne or corn. Its nest, according to Dubois, resembles much that of the Crested Lark; it is made of blades of grass and roots, lined with moss and root-fibres, sometimes with wool and feathers.

It builds twice a year, in April and June, and lays from four to six eggs, of a dirty white, covered with numerous spots of olive green, thickest generally at the larger end, though sometimes equally diffused. Long diameter one inch or nearly, short, three-quarters of an inch.

The adult male in breeding plumage has the upper parts rich brown, with the feathers bordered with russet. Inferior parts bluish white, with two large black spots on each side of the neck, forming a kind of half-collar, and separating the white throat from the russet and brown spots, with which the crop is mottled. Wing primaries blackish brown, the outer web very lightly bordered with white; the secondaries broadly tipped with that colour; under wing coverts and under part of primaries uniform blackish brown, relieved by the white shaft of the first quill, and the white tips of the secondaries. Two upper tail feathers brown, bordered with lighter, the two laterals white, the rest rich dark brown, tipped slightly with white; beak yellowish below and on the sides, dark brown along the upper ridge; feet yellowish brown; iris grey.

According to Degland, in the male after moult, or in autumn, the feathers above are darker in the centre, and their borders more russet. The plumage of the female resembles that of the male in autumn, but the head and beak are smaller, and the demi-collar in the neck is narrower.

The young after the first moult have the plumage

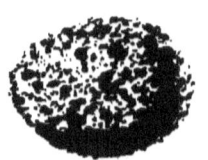

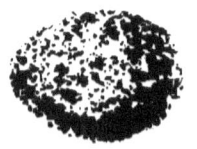

darker than adults, the upper feathers and those of the crop bordered with whitish; beak and feet yellowish.

Accidental varieties have been found white, spotted with white, grey or black, and others of an isabel colour.

My figure of this bird is from a male specimen sent me by Mr. Tristram, shot at Berronghina, May 28th., 1856. I have also selected for illustration two eggs from a series sent me by the same gentleman.

This bird has also been figured by Buffon, pl. enl. 363, f. 2; P. Roux, Ornith. Prov., pl. 185, f. 1, adult f. 2, head of young; Gould, B. of E., pl. 161; Bouteille, Ornith. du Dauph., pl. 30, f. 3; Vieillot, Faun. Franc., pl. 76, f. 1; Naumann, Naturg. Neue. Ausg., pl. 98, f. 1; Edwards, pl. 268; Dubois, Ois. de la Belg., part 61, pl. 102.

GRANIVORÆ.

Family ALAUDIDÆ.

Genus ALAUDA. *(Linnæus.)*

SIBERIAN LARK.

Alauda sibirica.

Alauda sibirica,	GMELIN; Syst., 1788.
" "	SCHLEGEL; 1844.
" "	DEGLAND; 1849.
" *calandra affinis,*	PALLAS; Iter. App., No. 15.
" " "	LATHAM; Ind.
" *leucoptera,*	PALLAS; Zoog., i, p. 518, No. 147, pl. 33, f. 2.
" "	BLAKISTON, in "Zoologist," vol. xv, p. 5509.
Phileremos sibirica.	KEYSERLING AND BLASIUS; Die Wirbelt, p. 37, 1840.
Melanocorypha leucoptera,	BONAPARTE.
Alouette de Siberie,	OF THE FRENCH.
Sibirische Lerche,	OF THE GERMANS.

Specific Characters.—First primary longest, but nearly equal to second; third about as much shorter than the second as the fifth is less than the fourth; the fourth seven tenths of an inch shorter than the third. Tail very narrow, the outer feathers white; hinder claw one fifth longer than the toe.

Length eight inches and one fifth; carpus to tip four inches and seven tenths; beak three fifths of an inch long by three tenths broad; hind claw half an inch to two fifths of an inch long.

This bird was thought by Pallas and Latham to be a variety of *A. calandra*. It is however a very distinct species, and the rounder form of the beak, the much slighter figure, the more pointed wing, and the difference in comparative length of the fourth primary, remove it altogether from that bird.

The Siberian Lark is an inhabitant of Siberia, Tartary, and Southern Russia, and rarely of Poland. It is also included by Captain Blakiston among the birds shot by him in the Crimea.—"Zoologist," 1857, p. 5509.

Its habits are described as similar to those of the rest of the family. By the kindness of Mr. Tristram, I am able to give a figure not only of the bird but its egg, both of which that gentleman received from Dr. Middendorff.

The bird has at first sight much the appearance of a Bunting. Captain Blakiston thus describes his meeting with it.—"Zoologist," 5509:—

"A few days after the 5th. of January I was again on the *qui vive*, as a friend told me he had seen some Buntings white below and rusty coloured above; with this hint I made for a camp, where he said some had been shot, the ground being covered with snow, and sure enough, on looking over a heap of small birds, I found the Calandra Lark, Common Bunting, and another new to me, which I put down for distinction as 'Lark Bunting, No. 20,' the skin as well as the sternum of which I preserved. The same officer a day or two afterwards kindly sent me a specimen of the same bird, the White-winged Lark, *(Alauda leucoptera,)* a male." This was determined afterwards by Mr. Gould, and Captain Blakiston gives a long and accurate description of it in its winter dress.

M. Ch. F. Dubois has an excellent figure of the bird, both in its young and adult plumage, with the following remarks:—"Though this bird is so rare in Europe, M. le Baron Selys-Longchamps possesses one, which was taken in the environs of Liege, in December, 1855. Having had it preserved, he kindly brought it to me to add to the supplementary list, before the family of Larks was concluded. The habits and propagation of this bird are very little known; its voice is not so agreeable as that of the Skylark, though its movements are equally elegant. It nests like it on the ground, in a slight excavation. It is not very timid, and allows people to approach it rather closely without fear."

A male specimen, sent me by Mr. Tristram from the Volga, without date, but from the freshness and brightness of the plumage, evidently in its nuptial robes, has the upper parts rich brown, bordered with russet, lighter on the nape; the top of the head, lesser wing coverts, and upper tail coverts, a brilliant red russet, which gives the bird a marked and distinctive character. The inferior parts are of a bluish white, with here and there a russet feather; the throat, crop, and sides of the neck spotted with brown and russet, the latter colour pervading the ear coverts. Under wing coverts and secondaries pure white, and the primaries blackish brown below; above, the primaries and secondaries are dark brown, the latter at first white on the inner web, becoming nearly entirely so in the middle. Tail feathers brown, with more or less white on their inner webs; the laterals quite of that colour. Beak livid, the upper mandible darkest; tarsi russet; feet dark brown.

My illustrations of the bird and its egg are from specimens sent me by Mr. Tristram; the former is a male. They were obtained from the keeper of the Imperial Museum of St. Petersburg, and are stated to be from Dr. Middendorff.

It has been also figured by Pallas, as *Alauda leucop-*

tera, and by Dubois, in his "Oiseaux de la Belgique," part 74, pl. 102—B, a very good figure of the adult male and the young.

This is the last of the Larks which I intend to introduce into this work, but there are one or two which merit a few words.

1.—*Alauda cantarella*, Bonaparte.—M. Dubois has given a good figure of this bird, the first if not the only one. It is smaller than the Skylark, and differs from it in some parts of its plumage. It was described by Bonaparte as distinct in Faun. Ital. Dubois says it is found in Italy in considerable numbers, and his doubts about it are completely removed. It is probably a race of the Skylark. I must, however, refer to M. Dubois' work for further remarks. Schlegel (Revue, 75,) says its specific distinction is not perfectly made out.

2.—*A penicillata*, Gould; *Phileremos scriba*, Brehm.—Of this bird, said to have been captured in the south-east of Europe, I cannot speak of my own knowledge.

3.—*A. bimacula*, Menetries; *Phileremos moreatica*, Mühle.—Is described by Keyserling and Blasius as a variety of the Short-toed Lark.

4.—*A. kollyi*, Temminck, pl. col. 305, f. i, and Man., 3, p. 202, is stated by Schlegel to be a doubtful species, allied to *A. brachydactyla;* and he is of the same opinion about—

5.—*A. pispoletta*, Pallas, Zoog. i, p. 526, No. 154.

6.—*A. deserti*, *Melanocorypha deserti*, Brehm, and *Annomanes deserti*, of Bonaparte and Lichtenstein, is said by M Dubois to have occurred accidentally in Europe.

END OF VOL. II.